W9-DFK-724

Manual of Bioassessment of Aquatic Sediment Quality

Edited by

Alena Mudroch
José M. Azcue
Paul Mudroch

Library of Congress Cataloging-in-Publication Data

Catalog information may be obtained from the Library of Congress

Lewis Publishers is an imprint of CRC Press LLC

No claim to original U.S. Government works
International Standard Book Number 1-56670-343-3
Printed in the United States of America 1 2 3 4 5 6 7 8 9 0
Printed on acid-free paper

Contents

Abbreviations

AES	Auger electron spectrometry
AES/AAS	atomic emission/absorption spectrometry
AMD	acid mine drainage
ANC	acid-neutralizing capacity
ANOVA	analysis of variance
APC	acid-producing capacity
APP	acid-producing potential
AQCS	Analytical Quality Control Services
ASC	sulfide-binding capacity
ASTM	American Society for Testing and Materials
AVS	acid-volatile sulfide
BACIP	before–after control–impact pairs
BCR	Community Bureau of Reference
BNC	base-neutralizing capacity
CANMET	Canada Centre for Mineral and Energy Technology
CCRMP	Canadian Certified Reference Materials Project
CDB	citrate–dithionate–bicarbonate
CEC	cation exchange capacity
CRM	certified reference material
CRS	chromium-reducible sulfide
DCB	dithionate/citrate/bicarbonate
DTPA	diethylenetriamine pentaacetic acid
EC	effective concentration
EDTA	ethylenediamine tetraacetic acid
GC	gas chromatography
IAEA	International Atomic Energy Agency
IBI	Index of Biotic Integrity
IGGE	Institute of Geophysical and Geochemical Exploration
IRMA	Institute of Rocks and Mineral Analysis
LOEC	lowest observed effect concentration
L/S	liquid/solid
MPN	most probable number
NBS	National Bureau of Standards
NH_3	ammonia

NH_4^+	ammonium ion
NIES	National Institute for Environmental Studies
NIST	National Institute of Standards and Technology
NO_2^-	nitrite
NO_3^-	nitrate
NOEC	no observed effect concentration
NRCC	National Research Council of Canada
NWRI	National Water Research Institute
OXC	oxidation capacity
PAH	polynuclear aromatic hydrocarbon
PCB	polychlorinated biphenyl
PCDD	polychlorinated dibenzo-*p*-dioxin
PCDF	polychlorinated dibenzofuran
PCR	polymerase chain reaction
ppb	ng/g
ppm	µg/g
ppq	fg/g
ppt	pg/g
QA/QC	quality assurance/quality control
RAS	reducing agent supplement
REC	reduction capacity
RIAP	Research Institute of Applied Physics
RM	reference material
SACCRM	South African Committee for Certified Reference Materials
SDS	sodium dodecyl sulfate
SEM	simultaneously extracted metal
SEM/XRM	scanning electron microscopy/X-ray microanalysis
SEP	sequential extraction procedure
SIMS	secondary ion mass spectrometry
SQC	sediment quality criteria
SRB	sulfate-reducing bacteria
SRM	Standard Reference Material
TU	toxic unit
USGS	U.S. Geological Survey
XPS	X-ray photoelectron spectrometry (10 to 30 Å)
XRPD	X-ray powder diffraction spectrometry

List of tables

List of figures

Preface

Interest in studying aquatic sediments has increased considerably during the last decade. At present, these sediments are considered an *in situ* source of pollutants in many aquatic ecosystems. Techniques used in physico-chemical analyses of the sediments were improved and new techniques for bioassessment of sediment quality have been developed.

Extensive survey, monitoring, and research activities, generally very expensive, are required to assess the extent and severity of sediment contamination, to evaluate the effects of contaminated sediments on aquatic ecosystems, and to prepare a plan for proper remedial action. These activities require sampling and different biological and physico-chemical analysis of bottom sediments. The use of incorrect methods in the analysis of sediments may lead to wasted money and human effort and to erroneous conclusions. A comprehensive monograph on physico-chemical methods used in the evaluation of sediment quality was recently published.* The objective of this book is to provide information on selected techniques for bioassessment of sediment quality. The bioassessment techniques are complementary to the physico-chemical analytical methods for quantitative determination of different elements and compounds used in the evaluation of sediment quality. The book contains a description of recently developed methods for bioassessment of sediment quality, including field and laboratory methods and methodologies used in studies of bacterial populations in the sediments. The book also contains a comprehensive review of chemical forms of different elements and compounds in sediments and evaluation of their availability to aquatic biota. In addition, discussions of the use of reference materials in sediment analysis and safety in the laboratory are included.

Alena Mudroch
José M. Azcue
Paul Mudroch

* Mudroch, A., Azcue, J.M., and Mudroch, P., Eds., *Manual of Physico-Chemical Analysis of Aquatic Sediments,* Lewis Publishers, Boca Raton, Florida, 1997, 287.

Acknowledgments

We wish to thank all the authors for their effort in the preparation of the chapters and for their excellent cooperation during editing of the manuscripts. Further, we would like to express our thanks to the many people who assisted in the preparation of this book. In particular, we are grateful to Ms. Angela K. Lee, National Water Research Institute (NWRI), Burlington, Ontario, for all the (re)typing and formatting of the manuscript, which she undertook with great skill, patience, and dedication. We would like to give special recognition to and greatly appreciate the editorial assistance of Mrs. Dianne Crabtree, also of the NWRI. The excellent contribution of the members of the Graphic Arts Unit of NWRI is also gratefully acknowledged.

Alena Mudroch
José M. Azcue
Paul Mudroch

Editors

Alena Mudroch, M.Sc., is an emeritus scientist with the National Water Research Institute, Environment Canada, Burlington, Ontario. Mrs. Mudroch graduated with a diploma from the Chemistry Department, State College, Prague, Czech Republic, and obtained her M.Sc. degree in 1974 from the Department of Geology, McMaster University, Hamilton, Ontario.

Mrs. Mudroch has published over 100 scientific papers and reports and has presented over 50 papers at national and international conferences and workshops. She is co-editor and co-author of the books *Handbook of Techniques for Aquatic Sediments Sampling*, published in 1991 by CRC Press, Boca Raton, Florida; *Handbook of Techniques for Aquatic Sediments Sampling*, second edition, published in 1994 by Lewis Publishers, Boca Raton, Florida; *Manual of Aquatic Sediment Sampling*, published in 1995 by Lewis Publishers, Boca Raton, Florida; and *Manual of Physico-Chemical Analysis of Aquatic Sediments*, published in 1997 by Lewis Publishers, Boca Raton, Florida. Her current major research interests include the characterization of aquatic sediments; defining the role of the pathways, fate, and effects of sediment-associated contaminants in aquatic ecosystems; and remediation of contaminated sediments.

José M. Azcue, Ph.D., is a research scientist with the National Hydrological Institute, Lisbon, Portugal. In 1981, Dr. Azcue graduated from the University of Basque Country in Bilbao, Spain, with a B.Sc. degree in biology and ecology. In 1983, he received a Technical Agronomist diploma from the Ministry of Agronomy, Spain. He obtained his M.Sc. degree in biophysics from the Federal University of Rio de Janeiro in 1987 and his Ph.D. degree in geochemistry from the University of Waterloo in 1992, having conducted his research on the mobility of arsenic in abandoned mine tailings.

During 1990, he lectured for the FURJ, where he co-organized an international course entitled "Sampling of Aquatic Environments." Among other publications, Dr. Azcue is co-author of the books *Metals en Sistemas Biologicos*, published in 1992 by the University of Barcelona, Spain, and *Manual of Aquatic Sediment Sampling*, published in 1995 by Lewis Publishers, Boca Raton, Florida, and co-editor of *Manual of Physico-Chemical Analysis of Aquatic Sediments*, published in 1997 by Lewis Publishers, Boca Raton, Florida. His current major research interests include geochemical cycling and analytical determination of contaminants in the environment, industrial and residential waste treatment, technology transfer, and environmental education.

Paul Mudroch, B.Sc., is a physical scientist with the Federal Programs Division, Environmental Protection Service, Environment Canada, Nepean, Ontario. Mr. Mudroch obtained his B.Sc. degree in geology and environmental studies in 1978 from the Department of Geology, University of Pennsylvania, Philadelphia.

His work experience includes assessment of contamination of the aquatic environment at abandoned mining sites and evaluation of the effects of oil and gas exploration on sediment quality in the Beaufort Sea. For four years he supervised analysis of different environmental materials in the laboratory of Indian Affairs and Northern Development, Yellowknife, Northwest Territories. His current activities include assessment of dredging projects at the federal facilities in Ontario, particularly the methods of sediment sampling and evaluation of sediment quality. He is co-editor of *Manual of Physico-Chemical Analysis of Aquatic Sediments,* published in 1997 by Lewis Publishers, Boca Raton, Florida.

Contributors

Venghuot F. Cheam is a research scientist with the National Water Research Institute, Environment Canada, Burlington, Ontario. Dr. Cheam graduated from the University of Oklahoma with a B.Sc. in chemistry in 1966 and a Ph.D. in chemistry in 1971. He has recently focused on research topics such as development of water and sediment certified reference materials, analytical methods, and the ultrasensitive laser-excited atomic fluorescence spectrometer and associated methods for the determination of trace metals in environmental samples including sediments. He has also worked extensively on the various aspects of quality assurance/quality control of the analytical measurement process.

Colette MacKenzie earned a B.Sc. in molecular biology and genetics from the University of Guelph in 1987. While working at the Ontario Veterinary College, she carried out research involving the genetics of *E. coli* and *Salmonella.* She is currently pursuing graduate studies at Northern Arizona University, specializing in the management of scientific research and technology.

Rudolf Reuther is a geochemist associated with the Swedish Environmental Research Group (MFG). He received his Ph.D. in environmental geochemistry from the University of Heidelberg in 1983, conducting a study on the chemical form of metals in lake sediments affected by acid precipitation from Sweden, Norway, and Canada. Since 1983, Dr. Reuther has gained both in-depth knowledge and broad international experience in chemical pollution control and monitoring. He has specialized in particular in the chemical speciation and ecotoxicology of priority metals, such as mercury and arsenic, and has published more than 30 papers, technical reports, and chapters in books. At present, he is mainly occupied with the implementation of effect analyses for industrial plans and projects, as part of environmental assessments. His clients originate from both the public and private sector and include national environmental protection agencies, regional water and waste bodies, the mining and metal industry, as well as international development agencies such as the World Bank. Dr. Reuther has been accredited as an external commissioner for solid waste management according to §11 of the German Waste Disposal Act.

Trefor B. Reynoldson worked for ten years with the Alberta Department of Environment in surface water quality monitoring. During that period, he was involved in establishing a biological monitoring network based on invertebrate community assessments. Dr. Reynoldson was also with the International Joint Commission for three years, during which time he was involved in the development of protocols and guidelines for the assessment of sediment contamination for the Great Lakes. For the past nine years, he has been a research scientist with Environment Canada, at the National Water Research Institute, where a primary research focus has been the development of site-specific biological objectives using benthic invertebrate community structure and toxicological end points for *in situ* sediment assessment.

Pilar Rodriguez is presently a titular professor of animal biology at the University of Basque Country (Spain), where she teaches zoology and invertebrates as indicators of water quality and toxicity in freshwater systems. She graduated in 1978 and earned her Ph.D. in biology in 1984. Dr. Rodriguez has 17 years of experience in freshwater benthic invertebrates, although her research has primarily focused on aquatic oligochaete systematics and pollution biology. Since 1989, she has been conducting research on industrial effluents and river sediment ecotoxicity.

Gordon Southam was educated at the University of Guelph, where he earned his B.Sc. (1986) and Ph.D. (1990) in microbiology. After completing his postdoctoral studies, he joined the faculty at Northern Arizona University in 1994 as an assistant professor of microbiology and environmental science. His research focuses on the bioremediation of acid mine drainage and fundamental biogeochemical processes, with a special interest in the formation of placer gold by bacteria. Dr. Southam is the author or co-author of 23 journal articles, 24 conference proceedings, and 6 book chapters.

chapter one

Trace metal speciation in aquatic sediments: methods, benefits, and limitations

Rudolf Reuther

1.1 Introduction

Global industrialization has significantly changed natural fluxes of elements, such as carbon, sulfur, nitrogen, and phosphorus, as well as those of toxic heavy metals. From the more than 90 elements in the periodic table, all but 20 are characterized as metals, with about 59 classified as heavy metals (i.e., density >5 g/cm^3). Seventeen of these metals are considered very toxic and available at toxic levels, and nine exceed the mobilization rate of natural geological processes (Novotny, 1995). For example, total anthropogenic releases of mercury to the atmosphere (i.e., 3500 tons/year) have reached approximately the same rate as natural emissions (i.e., 2500 tons/year) (Pacyna and Münch, 1991). As most elements escape from air and water by adsorbing to particulates, aquatic sediments are considered to be a major sink reflecting the current and past quality of elemental systems (Züllig, 1956).

Sediments are not well defined but rather are a complex heterogeneous mixture of different gaseous/liquid/solid, inorganic/organic, and living components derived from various sources and controlled by numerous physical, chemical, and biological variables. Sediments may also act as a source of contamination, when environmental conditions change. As carriers of particle-associated pollutants, aquatic sediments are used today to:

1. Identify, monitor, and control sources and the dissipation of metals
2. Study the transfer of pollutants between various ecosystem compartments

1-56670-343-3/99/$0.00+$.50

3. Assess the impact of metals in the environment
4. Prove the validity of control and remedial measures

The following is a brief overview of the present understanding of the occurrence of different forms of trace metals in sediments and its relevance to controlling their behavior and fate in aqueous systems. The definition and solution of this problem are presented in this chapter. Further, cornerstones of the development of methods for understanding the behavior and fate of metals are given, followed by a detailed description of currently available methods and their limitations. Finally, main mechanisms, processes, and factors with regard to metal binding, mobility, and bioavailability are discussed.

In general, there is still a lack of knowledge of the relationship between water and sediment quality and its significance for biological systems (Förstner, 1995). Sediments have long been neglected as a potential environmental scavenger for substances such as metals in surface waters. In contrast to organic micropollutants, such as PAHs and PCBs, trace metals are generally nondegradable. Although transformed by a variety of biogeochemical and physical processes in water, they will persist and primarily end up in bottom sediments. However, current quality criteria often refer to the water column as substrate for metal accumulation. Moreover, metal concentrations in sediments can adversely affect organisms at the sediment/water interface or in the sediment, despite the fact that their concentrations in water still comply with existing standards. From field and laboratory data, it is known that adverse effects of metals on biota occur even at concentrations of metals in sediments well below accepted threshold values (Chapman et al., 1992).

In some countries, guidelines for metal concentrations to assess the quality of soils and sediments are now being developed in order to estimate potentially severe effect levels (Table 1-1). These quality criteria are primarily based upon the equilibrium partitioning of metals in sediment pore water and sediments and relate to the content of organic matter/total organic carbon and known biological effects (Allen, 1993). However, the guidelines only express total concentrations of metals assuming that all chemical forms in which a metal may exist behave in a similar way (Martin et al., 1987). Today, it is well accepted that it is the particular chemical form of an element, rather than its total concentration, which rules its behavior according to prevailing physico-chemical and biological characteristics of the environment (Landner, 1987).

The necessity to analyze different chemical forms of a metal became evident after it was found that the toxicity of some metals, such as mercury and arsenic, strongly depends on their environmental speciation (Figure 1-1) (Bernhard et al., 1986).

With regard to sediments and other solids, the concept of *chemical speciation* has evolved to include various operational procedures to (1) deter-

Table 1-1 Quality Criteria for Metals in Soils and Sediments

Metals	mg/kg dry weight		
	Netherlands[a]	Canada[b]	Germany[c]
Chromium	250	111	100
Cobalt	50		
Nickel	100		50
Copper	100	114	100
Arsenic	30		
Molybdenum	40		
Cadmium	5		3
Mercury	2		2
Lead	150	250	100
Zinc	500	800	300

[a] B-value, Leidraad Bodemsanering, RVJM, Netherlands, 1983.

[b] Guidelines for the protection and management of aquatic sediment quality in Ontario, Ontario Ministry of the Environment, Ontario, Canada, 1992.

[c] German sewage sludge directive (AbfKlärV), 1982.

mine typical metal species (i.e., the form of binding between a metal and solid substrates) and (2) describe the distribution of metals among these forms and their transformation (Reuther, 1996). At the same time, the speciation of metals is an attempt to integrate both chemical (reactivity) and biological (availability) aspects of uptake and accumulation processes in sediments (Nelson and Donkin, 1985).

Physically, aquatic sediment systems can be divided into three main categories: (1) suspended (>0.45 μm), (2) bottom sediment, and (3) sediment

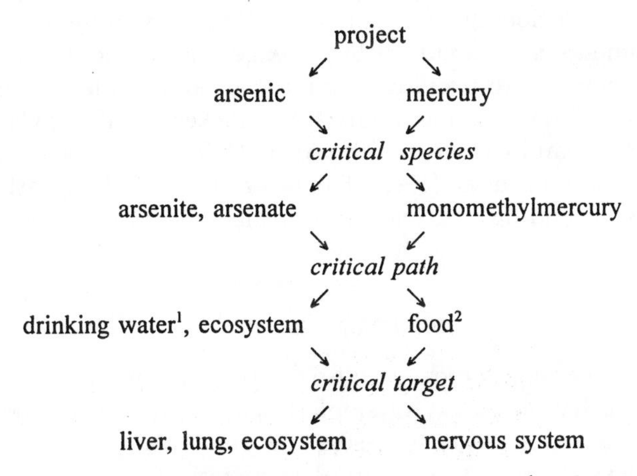

Figure 1-1 Critical forms and pathways of arsenic (50 μg/L)[1] and mercury (500 μg/kg),[2] WHO recommended values.

pore water. Each of these compartments contains a variety of individual physico-chemical forms of metals with mostly unknown transition states.

Generally, there are direct and indirect methods available to determine the speciation of metals in solid material. These include (1) instrumental methods and (2) chemical leaching and thermodynamic modeling. Direct instrumental analysis can be used to study the surface chemistry of particles, for example, by X-ray photoelectron spectrometry (XPS), scanning electron microscopy/X-ray microanalysis (SEM/XRM), secondary ion mass spectrometry (SIMS), and Auger electron spectrometry (AES) (Keyser et al., 1978). Leaching tests are grouped into equilibrium (e.g., batch and column tests, simpler and more reproducible) and diffusion-controlled methods (e.g., tank leaching test). Sequential extraction procedures are applied to estimate the relative strength of association between metals and various accumulative phases in the sediment (Martin et al., 1987). Geochemical numeric models use thermodynamic and kinetic data, such as stability constants, adsorption equilibria, and rate constants, to calculate and predict the dissolution behavior of distinct metal ions, molecules, and solid compounds in relation to major parameters. The large number of metals and ligands in natural systems requires the use of computer programs, such as MINEQL (Westall et al., 1976) or MINTEQA1 (Brown and Allison, 1987) to simplify the computational task. An excellent review of computer models for aqueous speciation was presented by Nordstrom et al. (1979). This chapter focuses on chemical extraction procedures as the most widely used approach to speciation of metals in sediments. For convenience, Table 1-2 provides a glossary of acronyms used in this chapter.

1.2 Literature review

A tremendous number of reports have been produced during the last 20 years on the partitioning of metals among various forms of particulates, such as sediments, soils, solid wastes, sewage sludge, and dredged material. Excellent reviews were provided by Förstner and Müller (1974), Förstner and Wittman (1981), Leschber et al. (1985), Pickering (1986), Campbell and Tessier (1987), Baudo et al. (1990), Allen (1993), Salomons and Stigliani (1995), and Salomons et al. (1995). However, the underlying principles and mechanisms of binding and release of metals in solid materials are not yet fully understood.

1.2.1 Historical development of chemical extraction procedures

Chemical extraction schemes were first applied in agronomy and exploration geochemistry. A variety of single leaching procedures were evolved, particularly in soil science, to assess the plant-available portion of trace nutrients (Jackson, 1958). Geochemical characterization of stream sediments has been standard practice in mineral exploration since the 1960s (Hawkes

Table 1-2 Glossary of Acronyms Used in This Chapter

Acronym	Definition
AES	Auger electron spectrometry
AES/AAS	Atomic emission/absorption spectrometry
ANC	Acid-neutralizing capacity
APC	Acid-producing capacity
APP	Acid-producing potential
ASC	Sulfide-binding capacity
AVS	Acid-volatile sulfide
BNC	Base-neutralizing capacity
CEC	Cation exchange capacity
CRS	Chromium-reducible sulfide
DCB	Dithionate/citrate/bicarbonate
DTPA	Diethylenetriamine pentaacetic acid
E°	Standard-state electrical potential of a redox reaction
EDTA	Ethylenediamine tetraacetic acid
L/S	Liquid/solid
Me^{2+}	Two-valent metal
OXC	Oxidation capacity
REC	Reduction capacity
SDS	Sodium dodecyl sulfate
SEM	Simultaneously extracted metal
SEM/XRM	Scanning electron microscopy/X-ray microanalysis
SEP	Sequential extraction procedure
SIMS	Secondary ion mass spectrometry
SQC	Sediment quality criteria
XPS	X-ray photoelectron spectrometry
XRPD	X-ray powder diffraction spectrometry

and Webb, 1962). Exploration geochemists tried to identify anomalies in the natural distribution of chemical elements within a given catchment area. In particular, streambed sediments have been used to indicate the composition of upstream rocks or zones of increased mineralization and weathering of surficial ore deposits. For example, loosely bound metals in stream sediments were used as an indicator for the existence of sulfide oxidation upstream (Rose, 1975). For sediment petrographic studies, it was important to define authigenic and detrital sedimentary phases. In general, a wide range of extracting agents have been examined, encompassing all relevant modes of chemical attack of metals associated with different phases (Pickering, 1986). Environmental geochemists adapted the methodology and integrating capacity of sediments in order to detect both current and past contamination events (Allan, 1974). The use of dated sediment cores and spatial concentration profiles allowed local background values and sources of contamination to be established.

Goldberg (1954) considered different sources of metals in particulates, such as biogenic, lithogenic, hydrogenic, cosmogenic, and atmospheric, by using reagents with different extracting properties (for example, weak acids, chelating and reducing agents, etc.). Goldberg and Arrhenius (1958) used ethylenediamine tetraacetic acid (EDTA) to differentiate between nonresidual (authigenic) and residual (lithogenic) fractions of metals in a pelagic sediment. Chester and Hughes (1967) tested three different extractants (i.e., NH_2OH, HCl, and CH_3OAc) to distinguish between carbonate-bound, Fe- and Mn-associated, and adsorbed trace metals. Ten years later, Agemian and Chau (1977) and Malo (1977) considered the use of 0.5 *M* HCl, or CH_3COOH, and 0.3 *M* HCl to estimate proportions of exchangeable metals and metals associated with carbonate and easily soluble Fe and Mn oxides or attached to organic matter. Today, single extraction steps are designed to recover individual (i.e., mainly exchangeable) metal fractions by cations or anions, which are more strongly bound to exchange positions (e.g., by use of $BaCl_2$, $MgCl_2$, or NH_4OAc), and organically bound metal species by using oxidizing agents, dissolution, proton displacement, and chemical competition (Salomons and Förstner, 1980). These simple one-step extractions proved rather successful in differentiating between anthropogenic and natural background levels. However, due to unknown reaction kinetics, only operationally defined procedures are obtained (Bruemmer et al., 1986).

In contrast, sequential extraction procedures were developed which include the successive leaching of metals from the sediment. These procedures may provide more information on the relative strength of major metal-binding components.

Salomons and Förstner (1980) divided the total metal content of sediments into three categories: (1) exchangeable-phase metals, carbonate- and easily reducible-phase (i.e., ferromanganese hydrous oxides) metals extractable by acid NH_2OH–HCl, (2) sulfidic- and organic-phase metals extractable with acidified H_2O_2 and reextracted in NH_4OAc (to inhibit readsorption), and (3) residual-phase metals, recovered by strong acid digestion ($HF/HClO_4$).

Robins et al. (1984) used the following four operationally defined fractions: (1) carbonate-associated metals recovered by an HOAc/NaOAc mixture (pH 5), (2) organically bound metals recovered by hot $NaHCO_3$ (pH 9.2) with 1% sodium dodecyl sulfate (SDS), (3) metals associated with ferromanganese oxyhydroxides recovered by NH_2OH–HCl, in Na-citrate (pH 5), and (4) inert bound metals recovered by an HF/HNO_3 digestion.

1.2.2 Recent attempts

Today, increasing analytical selectivity and accuracy make it possible to detect elements extracted from particulates at very low concentrations (Byrdy and Caruso, 1994). Most chemical extraction techniques are based on the "Tessier method" (Tessier et al., 1979) and differ only in practical details,

Table 1-3 Examples of Sequential Extraction Schemes

Fraction	Phase speciation[a]	EC/BCR version[b]	Tessier method[c]
Exchangeable ions		0.11 mol/L acetic acid	1 *M* NaOAc, pH 8.2
Carbonate-bound metals			1 *M* NH_2OH HCl in 0.01 *M* HNO_3, pH 2
Easily reducible phase	0.25 *M* NH_2OH HCl in 25% HOAc	0.1 *M* NH_2OH HCl in HNO_3, pH 2	0.1 *M* NH_2OH HCl 0.01 *M* HNO_3, pH 2
Moderately reducible phase			0.02 *M* $(NH_4)_2C_2O_4$ in 0.2 *M* $H_2C_2O_4$, pH 3
Sulfidic/ organic phase	0.1 *M* $Na_4P_2O_7$ (organic matter)	(2×) 8.8 *M* H_2O_2, reextracted in NH_4OAc/HOAc	30% H_2O_2, pH 2 by HNO_3, reextracted by NH_4OAc, pH 2 by HNO_3
	2× $KClO_4$ and HCl (resistant sulfides)		
Residual	HNO_3/$HClO_4$/HF	Lithogenic material	*Aqua regia* (HCl/HNO_3 3:1)

[a] Moore et al., 1988.

[b] Standard sequence by the Community Bureau of Reference (Ure et al., 1993).

[c] Tessier et al., 1979.

such as sample weight–extractant volume ratio, extraction time, washing between individual extraction steps, etc. (Table 1-3). Increasingly strong reagents are used to release metals associated with (1) exchangeable phase, (2) carbonate phase, (3) Fe/Mn oxides or reducible phase, (4) organic/sulfidic or oxidizable phase, and (5) residual mineral phase (Martin et al., 1987). These metal–solid associations are operationally defined and do not reflect actual forms of particulates or metals. The elution medium is used to simulate certain environmental conditions (Bernhard et al., 1986).

Obermann and Cremer (1992) recently developed a standardized leaching procedure at constant pH values (pH_{STAT} test) to simulate worst-case leaching situations, such as those that occur in landfills. Acid/base titrations (at constant pH 4 and 11) give a good estimate of the acid-neutralizing capacity of solids. In addition, a quantitative evaluation of the mobilizable part of metals is possible, in contrast to sequential procedures. By increasing or decreasing the liquid/solid (L/S) ratio ("cascade mode"), it is possible to assess the maximum leachability (high L/S) or the maximal possible leachate concentration (low L/S). The pH_{STAT} test is thought to complement existing standardized leaching tests, such as the German DEV-S4 standard test, which is more descriptive of the initial release of metals than long-term leachability.

The concentration of metals in sediment pore water is a highly sensitive indicator of reactions between metals bound to solids and the surrounding aqueous phase, particularly in fine-grained sediments with a large surface area. For this reason, determining the equilibrium partitioning of metals between sediment and pore water has been recommended as a criterion to assess the quality of sediments in relation to biota concentrations and biological effects (Di Toro et al., 1991).

A rather new approach is the acid-volatile sulfide (AVS) concept. The concept assumes that metal concentrations in the pore water of anoxic sediments are controlled by sulfides, which readily react with added metals to form highly insoluble metal sulfides. The added metals displace the Fe^{2+} from iron monosulfides (FeS) according to:

$$Me^{2+} + FeS = Fe^{2+} + MeS$$

The equilibrium is based on the lower solubility of transition metals compared with FeS. If all FeS is converted into MeS, then additional metals remain in solution. Typical concentrations of AVS sediment concentrations vary between 1 and 20 μmol/g sediment dry weight. Analytically, it has been shown that FeS does not originate from pyrite (FeS_2) (Di Toro et al., 1991; Allen, 1993).

Stability constants have been calculated for metal ions that form complexes with inorganic ligands, such as OH^-, HCO_3^-, NH_3, and to a lesser extent for those that form complexes with organic ligands, such as humic and fulvic acids. If the stability constants for these complexes are known, the distribution of a metal among these complex compounds can be established by thermodynamic principles, for example, by means of geochemical computer programs, such as MINEQL, MINTEQ2, or FITEQL (Westall, 1987; Eighmy et al., 1995). These computerized geochemical models are used to study the effect of solid surfaces on the dissolution behavior of metals. Stability constants for a variety of naturally occurring complexing agents and added metals were calculated. However, the lack of reliable thermodynamic data for natural organic materials limits their application in determining dissolved metal speciation (Table 1-4) (Neubecker and Allen, 1983). Metals bound to humic ligands were obtained by the lanthanide ion probe spectrometric method (Dobbs et al., 1989).

1.3 *Description of methods*

The ultimate goal of metal speciation described below is to isolate particular fractions of total metal content in sediments.

Table 1-4 Factors of Uncertainty Limiting the Thermodynamic Modeling of Solid Geochemical Speciation

1. Adsorption characteristics are related not only to system conditions (i.e., type and concentration of adsorbent) but also to changes in the net system surface properties resulting from particle/particle interactions
2. Effects of competition of various sorption sites
3. Reaction kinetics of individual constituents can hardly be evaluated in the complex and heterogenous sediment mixture
4. Definition of redox states
5. Characterization of natural surfaces and organic substance
6. Lack of reliable stability constants for metal–humic complexes
7. The small mineral database often available and the large number of iterations to approach equilibrium solution
8. Corrections of activity coefficients for high ionic strength
9. Complex ternary reactions
10. Assumed reversibility

After Bourg, 1995; Förstner, 1995.

1.3.1 *Sampling and sample preparation*

One has to note at the outset that, even with great care, sampling variance can be up to an order of magnitude greater than that of the analytical procedure (Pickering, 1986). Anoxic sediments should be transferred immediately to the laboratory and stored in acid-washed polyethylene containers in an inert and dark atmosphere at 4°C. Subsequent processing of the sediment samples should be carried out in a glove box under a nitrogen or hydrogen–nitrogen (1:19) atmosphere (Kong and Liu, 1995). Undisturbed surface sediments can be collected by grab samplers (e.g., Van-Veen or Petersen grabs), dredges (e.g., Ekman or Ponar dredge), or by different methods using SCUBA diving. Short sediment cores can be recovered by gravity corers or by a box corer when large sediment volumes are needed (e.g., for pore water extraction or benthic organism studies). Techniques for sediment sampling were described recently in detail by Mudroch and MacKnight (1994) and Mudroch and Azcue (1995). In addition to leaching, centrifugation, and squeezing, sediment pore water can be extracted *in situ*, for example, by the use of "peepers" (*in situ* diffusion pore water samplers) or dialysis bags which allow direct recovery of the dissolved metal fraction (Hesslein, 1976; Mayer, 1976; Adams, 1994; Mudroch and Azcue, 1995). Suspended matter (i.e., particles >0.45 μm) can be separated by high-pressure filtration, continuous-flow centrifugation, or bottom sediment traps (Håkanson, 1976; Müller et al., 1976; Rosa et al., 1994). Sediments may be further separated into fractions according to grain size (i.e., clay <2 μm, silt 2 to 63 μm, sand >63 μm), density (i.e., organics, organic–mineral, light and heavy minerals, by tetrabromethane 2.95 g/cm^3), or magnetic properties (particles <300 μm) (Müller, 1964; Maienthal and Becker, 1976; Bruemmer et al., 1986). In order to eliminate grain-size effects on concentrations of metals in sediments, extraction of the <63-μm particle fraction is recommended.

For anoxic sediments, all drying methods, such as air-, freeze-, and oven-drying, should be avoided, as they will inevitably change the original partitioning of metals (i.e., reduced organic and sulfidic phases) from more stable to more labile binding sites (except for Fe) (Kersten and Förstner, 1987). Drying may also cause aging effects by, for example, oxide crystallization, inducing a shift from carbonate to oxide phases. For extraction and storage of the extracts, acid-washed polycarbonate and polyethylene tubes or containers should be used. Quality control should include the use of duplicates, blanks, and certified reference materials in the running analytical routine (see Chapter 5 of this book for more information on using certified reference materials). The use of spikes and a cross-check with other laboratories may ensure a precision (reproducibility) and accuracy (recovery) of <15%, a level which seems sufficient for environmental samples (Tessier et al., 1979).

1.3.2 Direct measurements

A variety of X-ray and spectrophotometric analytical methods have been adapted to analyze the heterogenous composition of solid materials. Direct measurement of the metal concentration of different sediment components by electron microprobe or scanning electron microscope with an energy-dispersive system (SEM-XRM) allows the analysis of discrete metal-binding forms and probably gives the most reliable information on solid speciation (Bruemmer et al., 1986). Initial (im)mobilization effects on the surface of particles and depth profiles below the particle surface can be studied by electron microprobe X-ray emission spectrometry, electron spectroscopy for chemical analysis, AES, and SIMS (e.g., surface enrichment of Zn on mackinawite [FeS] by XPS) (Norsih, 1975; Eighmy et al., 1995). Depth profiles of Pb in calcite by SIMS indicated surface enrichment of adsorbed rather than coprecipitated species (Kheboian and Bauer, 1987). Sequential filtration, grain-size separation, density, and magnetic separation can be combined with direct measurement of particulate metals and surface composition by SEM-XRM. The forms of metal-binding solids can be further characterized by standard mineralogical procedures, such as X-ray powder diffraction spectrometry (XRPD), thin-section/light microscopy, or microprobe analysis (e.g., Müller, 1964; Jenkins and de Vries, 1970).

Charge densities on solid particles can be assessed either by the consumption of H^+ and OH^- ions by the sediment substrate in an acid/base titration or directly by a particle charge detector. The surface area of sediment particles is obtained by inert gas adsorption (Huang, 1981). Loss on ignition is a rough estimate of the total organic or volatile matter content of the sediment and can be obtained by dry combustion of the sample at 400 to 500°C for 3 to 6 hr.

Sediment pH and Eh can be measured directly by glass probe pH–Eh electrode.

1.3.3 Indirect methods

1.3.3.1 Total concentration of metals

The total metal content in sediments can be determined directly by X-ray fluorescence spectrometry or neutron activation analysis using powdered samples or indirectly by spectrophotometric determination (e.g., atomic emission/absorption spectrometry [AES/AAS]) after acid digestion. Acid treatments include both single acids (e.g., 5 *M* HNO_3 at 200°C or HF) and various acid mixtures designed to destruct the sediment (e.g., *aqua regia* [HCl/HNO_3 3:1]; H_2SO_4/K_2SO_4 or $HNO_3/KMnO_4$ for total Hg in Teflon bombs, with a preconcentration on Au-coated silica wool; or H_2SO_4/HNO_3 for total As, HNO_3/H_2O_2 for total Cd, HF/HNO_3 of ashed samples, $HNO_3/HClO_4$ 10:1, $HNO_3/HClO_4/HF$ 10:3:5, and HNO_3/HCl 3:1 in Teflon bombs). For complete digestion of refractory sediment components (Si, Al), digestion with HF or fusion with lithium tetraborate in Teflon crucibles or by microwave is recommended. The total recovery of only environmentally relevant metals may be sufficient by digesting with a concentration of HNO_3 at 170°C (Maxwell, 1968; Koch and Koch-Dedic, 1974; Breder, 1982; Tay et al., 1992). Detailed information on the digestion of sediment samples and techniques for determination of the concentrations of total metals was discussed by Hall (1996).

Quantification of dissolved metal concentrations (i.e., in pore water or sediment extracts) is mainly carried out by AES/AAS, or by inductively coupled plasma mass spectrometry, after filtration, and probably preconcentration steps. Minerals in the sediment can be characterized by XRPD, SEM, infrared analysis, thermo-analysis, and nuclear magnetic resonance spectrometry.

1.3.3.2 Binding capacity

Prognosis of the middle- to long-term behavior of metals is based on the sediment capacity to buffer protons and electrons. In sulfidic sediments, total pyritic sulfur (Sobek et al., 1978) or total sulfur (Bruynesteyn and Hackl, 1984) gives an estimate of the total acid- (proton) producing potential (APP). The addition of HCl to the sample, heating, and titrating with NaOH to pH 7, or with H_2SO_4 to pH 3.5, can provide an estimate of the acid- (proton) neutralizing capacity (ANC) in the sediment. The acid-producing capacity (APC) is calculated by subtracting APP from ANC. The "effective" APC can be easily defined as

$$APC = v/w \times H_e^+ - H_o^+$$

where H_o^+ = before and H_e^+ = after oxidation. For example, an $APC_{EFFECTIVE}$ of 5.12 mmol/kg was calculated for anoxic sediment from the River Elbe (Förstner, 1995). Simultaneous measurement of released metals may give an estimate of the potential of metal mobility in the sediment.

Another estimate of the buffering capacity of the sediment was suggested by Förstner et al. (1989a). Short-term metal mobilization occurs if the buffering capacity, expressed as Ca^{2+}, is low and the APP (S^{2-}) high. Both controlling parameters (i.e., ANC and APP) can be roughly estimated by extracting the sediment with acid NaOAc (for Ca^{2+}) and H_2O_2 (sulfidic/organic phase), as described by the method of Tessier et al. (1979). To predict the buffering capacity of anoxic solids, an acidity index has been introduced based on the concentration ratio between Ca and S in the sediment: Iac = lg $2c_{Ca}/3c_S = \Delta$ pH (Ca concentration as a measure of buffering capacity, S concentration as a measure of releasable acid, FeS oxidation releases $3H^+$, $CaCO_3$ can buffer $2H^+$; therefore Ca/S concentration ratio can serve as an approximation of the potential acidity of an anoxic sediment). Negative values of Iac will describe increasing acidity, while positive indices may suggest that the buffering capacity is sufficiently high (Förstner et al., 1989a). Operationally defined acid/base-neutralizing capacities (ANC_{24})/(BNC_{24}), as a measure to predict the mobility potential of solid systems, have been determined by the pH_{STAT} test (Obermann and Cremer, 1992).

The sulfide-binding capacity (ASC) can be derived by using the NH_2OH–HCl-extractable Fe concentration. The sum of the APC and ASC gives the maximum APC (for more details, see Förstner, 1995).

The redox buffering capacity in the sediment can be expressed by the total oxidation (OXC) and reduction capacity (REC):

$$OXC = 4O_2 + 5NO_3^- + 2Mn(IV) + Fe(III) + 8SO_4^{2-} + 4 \text{ (oxidized C)}$$

$$REC = 8NH_4^+ + Fe(II) + 2Mn(II) + 8S^{2-} + 4 \text{ (reduced C)}$$

The OXC can be related to Fe oxides and hydroxides and is determined by a 0.008 *M* Ti(III)–0.05 *M* EDTA extraction, followed by dichromate titration and AAS analysis. The REC corresponds to the chemical oxygen demand, which is determined by Cr_2O_7 oxidation and consists mainly of all Fe(II), reduced sulfur compounds, and oxidized organic matter in the sediment (Heron and Christensen, 1995).

AVSs are extracted by the cold acid purge and trap technique (i.e., converting S^{2-} into H_2S by dilute HCl [1 *N*, 3 *N*, 6 *N*]). The H_2S gas is purged by N_2 and trapped either in (1) NaOH or Zn acetate (colorimetric determination with methylene blue at 670 or 620 nm), (2) $Ag(NO_3)_2$ (determination by gravimetric difference of filter weights), (3) sulfide antioxidant buffer

Table 1-5 Forms of Extractable Sulfide in the Sediment

Fraction	Forms	Method
AVS	Amorphous iron monosulfides, soluble metal sulfides	Cold acid purge and trap technique
CRS	Biologically inert metal sulfides (e.g., pyrite FeS_2 and elemental S°)	Heating and adding $CrCl_2$
Heat-volatile sulfides	More reactive sulfides	Water bath at 95°C
Hydrogen sulfide (H_2S)	Molecular, volatile	N_2 purging, sulfide ion-selective electrode
Pyrite (FeS_2)		Pyrite = AVS – CRS

After Brouwer and Murphy, 1995.

(determination by sulfide-selective electrode calibrated with NaS_2), or (4) in a column immersed in liquid N_2 by photoionization detection (Casas and Crecelius, 1994). The simultaneously extracted part of metals (SEM) represents the portion which is released during AVS dissolution (Brumbaugh et al., 1994). In addition to AVS, chromium-reducible sulfides (CRS = FeS_2, metal sulfides, mainly amorphous FeS, and S°) and heat-volatile sulfides can also be assessed (pyrite = CRS – AVS) (Table 1-5) (Martin et al., 1996).

1.3.3.3 Partitioning factors

As a first approach to assess the strength of the metal–sediment bond, the ratio of the metal concentration in the solid and the dissolved phase can be assessed. The distribution coefficient is determined from laboratory experiments and defined as $k_D = Me_{SOLID}/Me_{DISSOLVED}$. The k_D values are related to the organic substance and other sorption-active surfaces in the sediment (Honeyman and Santschi, 1988). Batch-equilibrium methods generate equilibrium partition coefficients (Orth et al., 1995). Published partitioning coefficients (kp values = $conc_{SED}/conc_{WATER}$) for heavy metals vary between 1000 and 5000 and may reach up to 55,000 (Saradin et al., 1995). A correlation of the k_D with a chemical periodicity parameter, like the ionization potential or electronegativity, was observed for trace elements which bound strongly to solid phases (Whitfield and Turner, 1983). However, due to methodological constraints, such as pretreatment and separation techniques, and competing effects, its use as a criterion for sediment quality still remains limited (Shea, 1988).

Metal ions in natural waters, such as sediment pore water, are distributed among various complexing inorganic and organic ligands. Their distribution can be calculated if one knows (1) the complex-forming stability constants, (2) the ligand concentration, and (3) the ionic strength. For a given pH and ionic strength, the stability of a metal–ligand complex is expressed by the conditional stability constant: $k_{ML} = c_{ML}/c_M c_L$. Conditional stability constants (log k) show a large variation in natural waters, ranging from 9.5 (for Fe^{3+} complexes) to 3.3 (for complexes of Co^{2+} and Mn^{2+} with fulvic and humic acids) (Jenne and Luoma, 1977). Various methods to determine the complexing capacity and conditional stability constants are based on the titration of water samples with a suitable metal ion. For example, the complexation capacity of pore water of naturally occurring complexing agents can be determined by titrating the sample with Cu^{2+} ions (Allen, 1993). The end point of the titration curve coincides with the appearance of free Cu ions and the consumption of Cu^{2+} is equal to the complexing capacity. Free metal ions can be measured by voltametric (i.e., differential pulse polarography and anodic stripping voltametry), ion exchange (e.g., Chelex, Dowex, MnO_2), ion-selective electrode, dialysis, or biological methods, such as the decrease of photosynthetic activity (Neubecker and Allen, 1983). However, there is still much uncertainty about the nature of organically coordinated metal species (e.g., metals complexed by humic or fulvic acids). In general, their binding strength will increase according to the Irving–Williams rule: Mg < Ca < Cd ca. Mn < Cu < Hg (Irving and Williams, 1948). Chelating resins, such as Chelex-100 with an iminodiacetate as a strong ligand, are also used to estimate relative stabilities of soluble metal complexes and to differentiate between free metal ions and soluble complexed forms (Hendrickson and Corey, 1983).

The octanol/water-partitioning coefficient (k_{OW}) reflects the partitioning of lipophilic metal compounds, such as organotin, organoarsenicals, or organomercurials, between water and an organic (biotic) phase. The coefficient strongly varies in natural waters due to varying pH and salinity. For example, the K_{OW} of monomethyl Hg (MeHg) species may differ by a factor of 20, mainly because only uncharged species, such as MeHgCl and MeHgOH, partition into 1-octanol. Good agreement was found between theoretically calculated and experimentally derived K_{OW} values. The lower $K_{OW(\text{MeHg–OH})}$ value (0.07) was explained by stronger H bonds between water and the OH group, in contrast to the $K_{OW(\text{MeHg–Cl})}$ value (1.7), which indicates stronger interactions between octanol–HC groups (i.e., in the sediment organic matter) and the Cl group (Faust, 1992).

Plant and Raiswell (1983) concluded that simple mineral solution equilibria determine and explain the concentration of major elements, but not the solution behavior of trace elements in aquatic sediments, which seems more complex due to co-precipitation, surface effects, and the interaction with organic matter. Equilibrium conditions may not be very common in natural waters, due to supersaturation of one or more constituents (Novotny, 1995). Adsorption phenomena between dissolved trace metals and reactive surfaces may explain the lack of equilibrium. Apart from pH, adsorption of trace metals depends on adsorbent and adsorbate concentrations (Figure

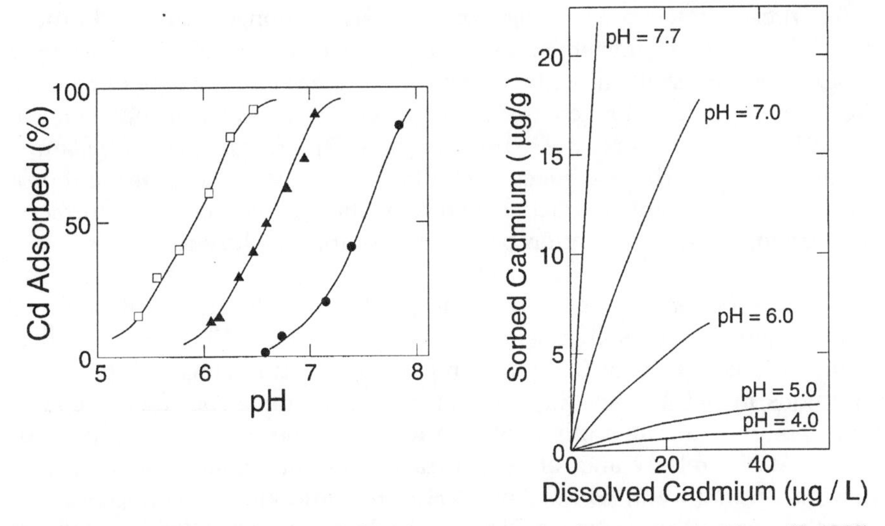

Figure 1-2 Sorption isotherms of Cd for various adsorbent concentrations ($Fe(OH)_3$: 1.3×10^{-3}, 1.3×10^{-4}, and 1.3×10^{-4} Fe mol/dm^3) and pH (Benjamin and Leckie, 1981; Christensen, 1984).

1-2) (Allard et al., 1987). Adsorption of dissolved organic compounds and metal complexes increases with the length of the alkyl chain, due to the growing hydrophobic nature. Also, adsorption depends strongly on the cation exchange capacity (CEC) of the sediment.

1.3.3.4 Chemical extraction

While thermodynamic calculations by using pH–Eh diagrams provide useful insights regarding the solubility behavior of principal solid–metal phases formed under specified conditions, the different modes of association existing between trace metals and major element compounds are much better resolved by empirical (i.e., analytical) approaches, such as chemical extraction. The distribution of metals among complexes formed with various ligands involves equilibrium as well as kinetic factors. When equilibrium calculations of thermodynamically stable oxidation states are compared with experimental or field data, it often turns out that it is the rate of reaction which determines the final form of association. For example, the complexity of processes affecting equilibria and rates of redox reactions is indicated by the relative stability of different redox states and by the fact that various redox couples coexist without approaching the equilibrium. Reduced forms of metals may occur in O_2-abundant waters, although the metals should exist in their highest oxidation state, according to the prevailing Eh determined by the O_2/H_2O redox couple (Burton and Statham, 1982).

Chemical extraction of sediments is intended to release metals associated with particular sediment phases. As reactions between solute metals and particle surfaces are heterogenous (i.e., adsorption, electron and proton transfers, etc.), extraction efficiency is more kinetically controlled and strongly depends on the availability of specific surface areas and the type of reactive sites (i.e., high- and low-energy sites). In general, results of the chemical extraction depend on (1) the extraction time, (2) L/S ratio, and (3) sample pH and buffer/carbonate effects. The pH seems most critical for the reproducibility of leaching/extraction experiments, as the charge of surfaces, adsorption, and the speciation of organic complexes depend on it.

1.3.3.4.1 Leaching tests. Leaching tests simulate the mobilization of metals from various solids, with the exception of the leaching time. It is assumed that increasing L/S ratios mimic long-term leaching effects, while variations in pH, Eh, and ligand composition emulate specific site conditions. Simple elutriate and percolation tests have been performed to simulate the initial release and long-term leachability of metals in solids under varying acidic (pH 3) or alkalinic (pH 12) conditions (e.g., in accordance with the German elutriate test DIN 38414). Extraction by water, in combination with ion-exchange resins, may simulate the leaching action of acid precipitation and uptake by biological surfaces. Treatment with water may recover salts and solutes entrapped in the sediment pore water, but along with a loss of exchangeable ions (depending on pH, temperature, and ionic strength). In addition, added surface-active agents may imitate natural surfactants, dispersing organic colloids. Low pH extraction may overlap with proton displacement and high pH extraction may result in a complex formation (more pronounced for acid-consuming components, such as $CaCO_3$ or hydrous oxides) (Hirner, 1996).

The pH_{STAT} test has recently been proposed as a standard leaching method to follow up the mobilization of metals under worst-case pH conditions (at constant pH 4 and 11). Another example is the elutriate test developed by the U.S. Environmental Protection Agency (U.S. EPA) to detect releases of metals in dredged sediments by mixing the material with water from the disposal site (at L/S = 4). If the concentrations of metals in the elutriate exceed those in the water at the disposal site by a factor of 1.5, the dredged sediments are considered unsuitable for open-water disposal (Förstner, 1995).

1.3.3.4.2 Single-step extraction. Single extraction procedures are often used in studies of uptake of trace elements by plants to estimate the availability of the trace elements. Metals can be recovered from the soil solution by a water saturation extract or by 0.1 *M* $CaCl_2$ or $Ca(NO_3)_2$ (Bruemmer et al., 1986). Soil fertility is commonly related to the EDTA-extractable metals in soils. Zinc extractable by NH_4OAc–dithizone or 1 *M* NH_4OAc (pH 4.6) is assumed to be plant available (although a part of Zn is also extracted from clays by this method) (Jenne and Luoma, 1977). For sediments, it is considered that available forms of metals can be extracted by 0.01 *M* EDTA (pH 8),

1 *M* NH_4OAc, or 0.5% ethanol (Chen and Morrison, 1991). Exchangeable K^+ and Ca^{2+} are extractable by an acid NH_4 lactate/HOAc buffer (pH 4), while total K and Ca are recoverable by dry ashing at 600°C and dissolution in 2 *M* HCl (Fawaris and Johanson, 1995). Ammonium nitrate can be used to estimate nonspecifically sorbed metals, while NH_4OAc is used to extract those specifically bound (Hirner, 1992). It was further suggested that carbonate minerals can be selectively extracted by an NaOAc buffer (Kersten and Förstner, 1986; Martin et al., 1996). However, there is no clear relationship between extractability and plant availability of metals and trace elements.

The U.S. EPA established a standard leaching procedure (Toxicity Characteristic Leaching Procedure) to estimate the potential toxicity of solid waste material by the use of a 0.1 *M* acetate buffer (pH 5) (U.S. EPA, 1986). A modified $BaCl_2$ extraction is used in Germany as a first estimate of the bioavailability of metals.

Extractions using dilute acids, such as 2.5 to 3.2% HOAc, 0.1 *M* HCl, and 2% H_2SO_4, are rather dissolution than diffusion controlled (Jenne and Luoma, 1977). Metals extractable by 0.5 to 3 *M* HCl are considered to be nondetrital (i.e., "bioavailable"), as most of the metals sorbed by oxides and organics, and part of the metals from resistant sulfides, become released (Brumbaugh et al., 1994; Brügmann, 1995).

The total concentration of Fe in sediments is a complex mixture of various Fe compounds, including amorphous and crystalline species, mainly oxides/hydroxides, carbonates, sulfides, and silicates. For example, extraction of amorphous or poorly crystalline Fe oxides (largely magnetite, partly hematite and geothite) may be released by acidic NH_4 oxalate or to a lesser extent by 0.1 *M* Na pyrophosphate or dithionate citrate. Total Fe(III) is recovered by 5 *M* HCl, while the use of 0.5 *M* HCl may provide a simple estimate of the Fe(II)/Fe(III) ratio and the redox potential in the sediment. In general, the presence of exchangeable Fe(II) may indicate the degree of reducing conditions in the sediment (Table 1-6). Dithionite is considered to extract total Fe oxides, but also Fe silicates, siderite, and part of Fe(II) sul-

Table 1-6 Main Extractable Forms of Iron in the Sediment

Fraction	Extraction
Exchangeable Fe(II)	Anaerobic 1 *M* $CaCl_2$
Total Fe(II)	5 *M* HClFe(II) + HI–Cr(II)HCl–Fe(II)
Total Fe(III)	5 *M* HCl
Total Fe(II)/Fe(III) redox estimate	0.5 *M* HCl
Fe(III) oxides (geothite FeOOH, hematite Fe_2O_3)	Ti(III)–EDTA[a]
AVS	Hot 6 *M* HCl
Pyrite (FeS_2)	Sequential HJ and Cr(II)–HCl

[a] Also gives an estimate of the oxidation capacity.

After Heron et al., 1994.

fides, while there might be an overestimate of the total Fe oxide when Fe(II) and Fe(III) are present. For this reason, a Ti(III)–EDTA extraction was suggested to estimate the Fe(III) oxide content, not including Fe in magnetite (Heron et al., 1994).

Oxidic Mn and Al components are extractable by acidic NH_2OH HCl or NH_4OAc + 0.2% hydroquinone, NH_4 oxalate, or pyrophosphate (Reuther and Grahn, 1985). Reactive silica dissolves in a dithionate citrate buffer or in 5% Na_2CO_3. In addition, the oxidation of organic matter by H_2O_2 dissolves part of Mn oxides and sulfides (Jenne and Luoma, 1977).

Humic acids can be extracted by 0.1 to 0.5 *M* NaOH (or 1 *M* ammonia) under an N_2 atmosphere, and precipitated by 6 *M* HCl (pH 2), while fulvic acids remain in solution (Griffith and Schnitzer, 1975; Luoma and Bryan, 1982).

1.3.3.4.3 Multistep sequential extraction procedures. Similar to single extractions, sequential methods measure no stoichiometrically defined metal species, but rather a group of species, or a fraction of the total metal content, as defined by the method (i.e., operationally defined). The methods usually identify a certain number of metal fractions (i.e., 3 or 7), which are dissolved by the use of progressively stronger reagents and by minimizing overlaps between the reagents. The least aggressive reagent is applied first. For example, H_2O_2 effectively solubilizes Mn oxides under mildly acid conditions. However, this step must come after the reducing step. This principle will always determine the sequence of the extractions. The extractants farther down in the extraction sequence reflect more long-term effects but with reduced prognostic accuracy. However, in contrast to single leaching, sequential extraction procedures (SEPs) make it possible to identify rearrangements of specific solid "phases" before actual remobilization of metals into the dissolved phase. In addition, valuable information on diagenetic postsedimentary processes can be obtained by SEPs (Martin et al., 1987; Förstner, 1995).

Most SEPs start to isolate water-soluble, exchangeable, and easily adsorbed cations (and anions) from the sediment, which may reflect and regulate short-term changes in the composition of the interstitial and overlying water column. This may be carried out by a water wash (i.e., yielding water-soluble ions and metal complexes) or by extraction using $MgCl_2$ or NH_4OAc.

The following is a short discussion of the main extraction steps in relation to associated reactions and conditions and their efficiency (Pickering, 1986). Ion displacement depends mainly on (1) the amount of adsorbed metals, (2) the weight of the tested solid material, and (3) the volume of the extractant. Because of different bonding constants (Pb, Cu, Cd, Zn > Ca > Mg >> Na, NH_4), H^+ may desorb more Me^{2+} than Mg^{2+}, and 1 *M* Mg^{2+} may desorb more Me^{2+} than 1 *M* Na^+, etc. Displacement of ions is enhanced by decreasing levels of adsorbate (i.e., by complexation with organic ligands, such as EDTA).

Mobilization of metals by salt anions is due to soluble complex or ion-pair formation and/or desorption of cations. Extraction efficiency of Cl^- salts varies with metal ions (Cd, Zn > Pb, Cu and $La^{3+} > Ca^{2+} > Na^+$) and with the type of substrate. Guy and Chakrabarti (1975) suggested the following general sequence of decreasing sorption capacity for metals: Mn oxides > humic substances > Fe oxyhydrates > clay minerals. It is assumed that the affinity for sorption sites has a greater effect than the formation of chloro complexes (0.05 *M* $CaCl_2$ > 1 *M* $MgCl_2$ > 0.5 *M* NaCl). Often, 1 *M* $MgCl_2$ is favored to extract the exchangeable fraction, but it may also affect specifically sorbed metals, while 5 *M* NH_4Cl also dissolves carbonates and partly organic matter.

Nitrate does not form stable metal ion complexes, while displacement occurs only through cationic competition. For example, 1 *M* NH_4NO_3 is used in studies of the uptake of metals by plants as a most sensitive indication, such as the uptake of Cd. However, NO_3 extractions are not used as often as Cl^- extractions.

The acetate anion forms slightly more stable complexes than chloride but interacts with Fe^{3+} and Ca^{2+}, leading to further dissolution of the sediment substrate and pH variations that affect buffering capacity. However, average recovery is considered to be about 10% higher than that in Cl and NO_3 extraction. For example, 1 *M* NaOAc (pH 8.2) is used to measure the CEC in soils, although it dissolves more carbonate than 1 *M* $MgCl_2$ (see above). Sodium acetate buffered at pH 5 recovers most of the carbonate-bound metals, but also releases specifically adsorbed metals. Ammonium acetate (1 *M*, pH 7) displaces exchangeable cations, but also attacks carbonates to a greater extent than 1 *M* NaOAc (pH 8.2) and partially attacks sulfates, oxides, and sulfides. Acetic acid (2.5%, pH 3) displaces exchangeable cations, but also more firmly bounds metals due to the low pH.

Reducing agents are mainly used to extract Fe and Mn oxides. The most widely applied reagents are NH_2OH–HCl (reduces Mn oxides) and oxalate (or dithionite) solutions (dissolving primarily Fe oxides). However, extraction efficiency depends on, in addition to pH and stability of the metal complex, the redox potential, which means that MnO_2 can be extracted by a reductant with a less positive E°. In addition to Mn oxyhydroxides, NH_2OH–HCl also dissolves small quantities of iron in the sediment, with decreasing pH. For example, NH_2OH–HCl in 25% HOAc leaches Mn and Fe oxides (and also affects calcite and smectite), and $(NH_4)_2C_2O_4$ (pH 3) dissolves mainly amorphous or poorly crystalline Fe and Al oxides, with lesser attack on silicates and crystalline geothite when used after the extraction of sediments by NH_2OH HCl. However, its efficiency varies with the degree of illumination. Amorphous oxides are mainly extracted in the dark, while crystalline oxides are extracted during daylight. A mixture of dithionite, citrate, and bicarbonate (DCB) has been used to dissolve crystalline and amorphous oxyhydroxides; however, it also attacks Fe-rich silicates. Lower recoveries of metals are observed due to the formation of insoluble S^{2-} caused by S_2O_4–S^{2-} disproportioning.

Generally, oxidizing agents cause the destruction of organic matter, such as living organisms, detritus, or organic coatings (Table 1-6). Oxidants should have a high positive redox potential, for example, $E°(Cl_2) = 1.36$ V or $E°(H_2O_2) = 1.77$ V. Oxidizing acids, such as HNO_3, $HClO_4$, HCl, or HF, dissolve organically bound metals as well as other mineral components, such as sulfides. In addition, fuming $HClO_4$ and HNO_3/HCl mixtures tend to react explosively. Similarly, NaOCl (buffered at pH 9.5), also a 30% H_2O_2 solution, acidified by dilute HNO_3, oxidizes the organic sediment fraction, with the following limitations: (1) no complete destruction of organic matter is achieved; (2) the reaction rate is slow; (3) formation of oxalate as a by-product, which desorbs, dissolves, and precipitates the metals; (4) MnO_2 dissolves at pH <5 and consumes the oxidant; (5) released metals are scavenged by Mn oxides; (6) sulfides are also oxidized; (7) the resulting acid solution may attack silicates; and (8) the colored extract may hamper spectrophotometric analysis.

The recovery of metals from organic matter, such as humic substances, may also occur due to (1) dissolution of organic acids by alkaline solutions, (2) conversion of polyvalent cations into complexes, or (3) the dispersion of colloids (i.e., humic acid polymers). Stable organic complexing agents have been employed to estimate the nondetrital part of metals in sediments. They reduce dissolved metal concentrations, but also dissolve part of the substrate by desorption processes. Only a small fraction exists as nonprotonated ligands. Metal–chelate equilibria are very pH and Eh dependent. EDTA and diethylenetriamine pentaacetic acid (DTPA) form very stable, water-soluble complexes with many polyvalent cations, but also dissolve part of the organic substrate. For example, Na_2–EDTA, buffered by 1 *M* NH_4OAc, specifically attacks amorphous oxyhydroxides. In addition, oxalate and citrate form complexes with most polyvalent cations, but with moderate stability. Pyrophosphate extracts metal–organic complexes by forming strong coordination complexes (chelates) (e.g., with Ca^{2+}, Mg^{2+}, divalent metals, and Fe^{3+} and Al^{3+}). For example, $K_4P_2O_7$ and $Na_4P_2O_7$ effectively remove metals bound to flocculated organic acids causing the dispersion of humates and fulvates. Compared with EDTA, pyrophosphates are considered to extract colloidal organics more efficiently, with less solution of oxides.

Other chelating organic solvents, such as chloroform, ether, gasoline, benzene, and carbon disulfide, isolate nonhumic organic sediment fractions, such as proteins, carbohydrates, peptides, fats, waxes, and resins. However, they are also used to preconcentrate low levels of dissolved metals, prior to analysis. The addition of surfactants, such as SDS, increases the amount of extracted organic matter, particularly in combination with strong bases or $Na_4P_2O_7$ (e.g., 0.1 *M* $Na_2B_4O_7$, pH 9.3, 1% SDS, or an $NaHCO_3$–SDS mixture, pH 9.2) (Hirner, 1992). Ion-exchange resins help to separate and identify organic metal substrates according to their relative bonding strength and due to the high adsorptive capacity (e.g., 1 to 5 eq/kg) of different functional groups which are specific for particular solute ions. For example, acid resins are suitable to dissolve carbonate minerals, while silica gel with sulfonic acid groups has a high CEC and a low affinity for organic compounds. The affinity between humic/fulvic acids and exchange resins is used to isolate organic acids, which can be further separated and characterized by gel permeation and reversed-phase high-pressure liquid chromatography.

Residual fractions consist mainly of refractory silicate or organic material and can be extracted by bomb or microwave digestion using strong mineral acids at elevated pressure and temperature.

Figure 1-3 represents the ability of various extractants to release metals retained in different modes of binding or by specific fractions in the sediment.

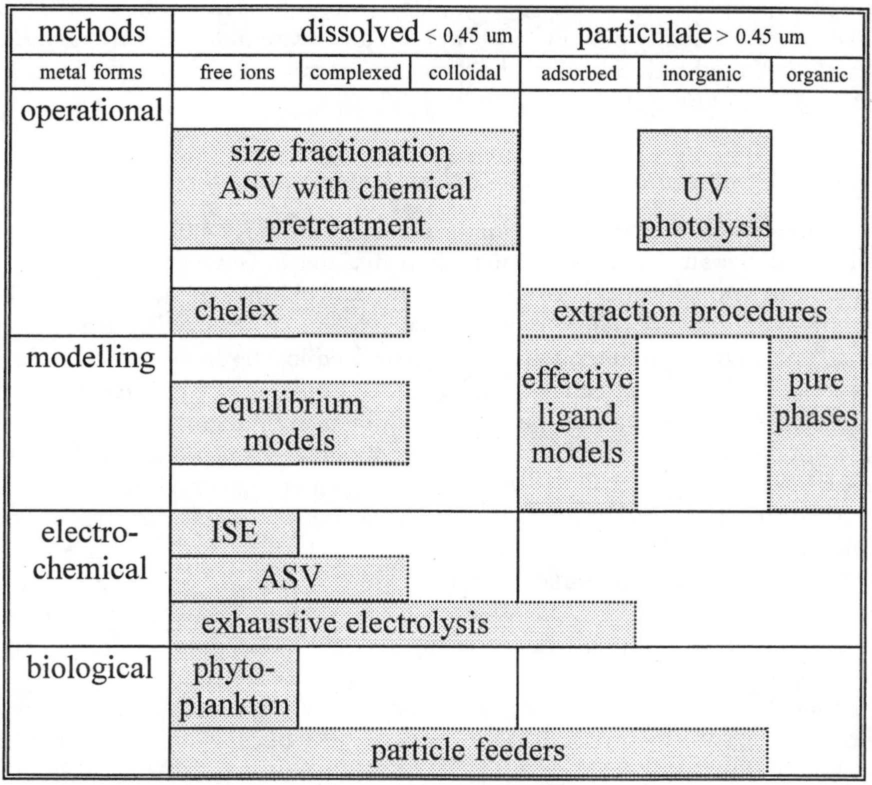

methods	dissolved < 0.45 um			particulate > 0.45 um		
metal forms	free ions	complexed	colloidal	adsorbed	inorganic	organic
operational	size fractionation ASV with chemical pretreatment				UV photolysis	
	chelex			extraction procedures		
modelling	equilibrium models			effective ligand models		pure phases
electro-chemical	ISE					
	ASV					
	exhaustive electrolysis					
biological	phyto-plankton					
	particle feeders					

Figure 1-3 General classification scheme for the ability of various extractants to release metals from various modes of binding in the sediment. (After Turner and Whitfield, 1980.)

1.3.4 Biological methods

Chemical speciation alone is not sufficient to assess or predict the biological effects of metals or possible synergistic effects among pollutants. It is known that different sediments with a similar quantity of metals behave toxicologically differently, mainly due to different types of metal binding and sediment characteristics. For example, the toxicity of many divalent metals in anaerobic sediments is controlled by variable amounts of AVS, but mainly by Fe and Mn oxides in aerobic sediments, in addition to organic substances. Bioassays have been successfully developed to indicate the occurrence of toxic metal species in (1) pore water, for example, by *Daphnia magna* (48 hr), rainbow trout (96 hr), and Microtox® (EC_{50}, 50% reduction of light output of luminescent bacteria, not acutely toxic for EC_{50} >100%), and (2) whole sediment, for example, by *Hyalella azteca* (28 days), *Chironomus riparius* (14 days), rainbow trout (21 to 28 days), and *D. magna* (2 to 22 days) (Kemble et al., 1994). In addition, direct sediment toxicity testing has been used to classify biological toxicity by using the following EC_{50} values (i.e., the percent of sample that produces an EC_{50}): $EC_{50} \leq 0.5\%$ = very toxic, $EC_{50} > 0.5\%$ and $< 1\%$ = moderately toxic, and $EC_{50} > 1\%$ = nontoxic. Compared with chemical extraction methods, these solid-phase bioassays are fast and inexpensive (Kwan and Dutka, 1995).

1.3.5 Combinations of biological and chemical methods

It is understood that a single approach alone will hardly elucidate the complex and dynamic nature of interaction that exists between the physical, chemical, and biological components and the binding of metals in sediments. Pure toxicity data will not explain the cause of bioaccumulation or food chain effects, while purely physical or chemical figures are not enough to describe the mechanisms of availability and uptake. The best results may be achieved by combining speciation with bioassay methods. For example, the uptake of Cu by worms correlated well with the concentration of easily reducible phases of Cu in the sediment, due to low redox and pH values in the worm guts, which caused the dissolution of Mn oxide coatings (Diks and Allen, 1983).

However, these combinations must include real-world parameters, such as realistic pH, dilution, and substrate concentrations, as well as sensitive and representative organisms or ecotoxicological end points (Orth et al., 1995). Bioassays involving sediment extracts are surely one approach to further close the missing link between leachability and bioavailability. However, the organisms are still only exposed to a portion of metals in the sediment (Brouwer and Murphy, 1995). In addition, the properties of single-metal associations are not additive (Bourg, 1995). Similarly, pore water tests address only water-soluble contaminants (Tay et al., 1992). Recently, the Microtox® solid-phase test was introduced to integrate the toxicity of soluble and insoluble metal fractions attached to organic or inorganic substrates in the sediment (Microbics Corporation, 1992). The combination of a direct method, such as surface-specific instrumental techniques (i.e., PIXE, SIMS, SEM/XRM), with indirect analytical methods, such as chemical extraction or titration, to assess the binding capacity of distinct surface mineral compounds seems very promising (Hirner and Xu, 1991).

The Sediment Quality Triad has been used as an effects-based approach to integrate data from whole-sediment laboratory toxicity exposures, benthic community structure (e.g., Chironomidae taxa richness), and chemical/physical analysis to (1) identify and differentiate pollution-degraded areas, (2) determine concentrations associated with effects, (3) predict where degradation may happen, and (4) rank areas for possible remediation (Chapman et al., 1992). However, this approach has also been used to develop numeric sediment criteria (Chapman et al., 1991). For example, sediments have been classified according to ranking variables, such as (1) body length and sexual maturation for the amphipod *H. azteca* (for sediment toxicity), (2) number of Chironomidae genera (for benthos impact), and (3) whole-sediment SEM and SEM/AVS (for sediment chemistry), which were scaled from 0 (lowest value) to 100% (highest value) (Canfield et al., 1994).

The status of bacterial communities living in the sediment is indicated by their enzyme activity, which can be inhibited by toxic forms of contaminants. Bioassays using sediment enzyme activities have been conducted in combination with chemical extraction using, for example, NH_4OAc and EDTA to trace the effect of storm water on urban rivers. It was demonstrated that low bacterial activities in the sediment coincide with high concentrations of extractable metals (Chen and Morrison, 1991).

Synthetic membranes have been used to measure the diffusive transport of bioavailable metals across biological barriers. In combination with chelating resins, such as Chelex-100, uptake rates can be estimated. However, active transport media, such as protein structures (e.g., metal transport into algal cells is considered to be slower), are not considered (Morrison et al., 1989). To study exchange reactions and uptake mechanisms of individual solid phases under changing system conditions, a multichamber device was used, where individual components, such as algal cell walls, Mn and Fe oxides, bentonite, and quartz, were separated by membranes which still allowed solid/solute interactions (Calmano et al., 1988).

A circulation system, together with an ion exchanger, was used to follow up the mobility of trace metals by the controlled increase of significant release parameters, such as pH, redox potential, and temperature. Together with the use of an SEP scheme before and after the experiment, extrapolation of the total mobilizable "metal pool" and thus the effect of long-term variations in the leachate composition was possible. Rearrangements of specific solid phases (i.e., their effect on metal mobilities) can be evaluated prior to metal mobilization. The truly mobilizable part of metals can be calculated by direct leachate concentration and by estimating mobilizable pools. A strong mobilization of Cd, in contrast to Zn and Cu, was observed at pH 5 and Eh +400 mV (Schoer and Förstner, 1987).

1.3.6 Limitations and problems of speciation methods

Chemical extraction techniques are limited particularly due to (1) changes in composition of metal species during sample handling and pretreatment, (2) nonselectivity of extractants, and (3) redistribution among phases during extraction (Kheboian and Bauer, 1987). Artificial partitioning processes may change the total concentration and species composition in both the dissolved and particulate phase (Förstner and Carstens, 1988). It is also difficult to compare the results of the extraction studies if different operational parameters, such as reagent concentration, sample composition, pH, L/S, temperature, extraction time, intensity, etc., were used.

1.3.6.1 Sample handling and treatment

Different procedures that occur during sampling and sample pretreatment, such as statistically invalid sampling, reactions with material of sample containers, sample contamination, pH changes, light-catalyzed reactions, freezing, lyophilization, evaporation, oxidation, biological activity, and different contact intensities between solids and solution (e.g., by using shakers, agitators, flow-through columns, dialysis bags), may be disruptive and change the original speciation of metals in the sediment, for example, from oxidizable to easily reducible or carbonate forms (Pickering, 1986; Kersten and Förstner, 1987).

There is no universally valid sample preservation technique available for all metals and in all fractions. Also, extraction processes, such as variable duration of extraction time, temperature, L/S, and changes in pH, may irreversibly alter the complexation state of metals (Martin et al., 1987). Losses of dissolved metals occur during filtration (e.g., due to Fe^{2+} oxidation and subsequent Fe precipitation), as well as by unwanted dispersion during particle size fractionation (Brumbaugh et al., 1994).

The appropriate extraction time depends mainly on the solid/solution ratio and the mixing intensity. The contact time between particles and extraction solution can be improved by ultrasonic treatment during which coatings are dislodged, particles disaggregated, and humic material partially separated. Equilibrium may be achieved within 12 to 24 hr, often with a typically rapid initial release of metals. However, the rate of many chemical reactions (1) doubles if the temperature increases by 10°C, (2) increases with increasing content of finer particles, and (3) changes according to the type of minerals involved (Pickering, 1986). Matrix problems due to strongly acid-consuming compounds, such as carbonates or metal oxides, will also hamper the leaching rate (Trefry and Metz, 1984). Incomplete and variable recoveries of spiked metals may also result from losses by washing solutions, due to manual manipulations, or due to the release of different elements from container walls when using strong extractants (Kheboian and Bauer, 1987). For spiked metals, possible substrate-induced redox reactions must be considered and separated from method-induced ones (James et al., 1995). Also, acidifying extracts may reduce the dissolved metal concentration due to precipitating humic substances according to the Irving–Williams sequence (Obermann and Cremer, 1992).

1.3.6.2 Oxidation effects

Aeration of anoxic sediments during storage will change the original fractionation pattern, with an increase in mobility of Ni, Pb, Cu, Zn, and Cd and a decrease in mobility of mainly Fe and Mn (Kersten and Förstner, 1987). Freeze-drying or the removal of water from sediments may oxidize sulfides or may lead to stable sulfide formation (e.g., FeS_2) (Brouwer and Murphy, 1995). In general, oxidation causes a shift to more easily extractable phases. As an example, oxidation caused the rearrangement of Cd toward weaker binding sites (i.e., a shift from strong organic surface complexes by SH binding to weaker organic and Fe oxide surface ligands), resulting in an erroneous determination of high bioavailability (Figure 1-4). About 44% of Cd originally bound to organic matter in an anoxic sediment was found in the "carbonate" fraction, after exposure to air (Martin et al., 1996). Rapin et al. (1983) showed that FeS in anoxic sediments was solubilized earlier during the sequential extraction when the tested sediment was extracted in air instead of an N_2 atmosphere. However, oxidation of labile sulfide minerals may even occur during extraction under an N_2 atmosphere (Kheboian and Bauer, 1987). Leonard et al. (1995) reported that significant oxidation of AVS in frozen, air-exposed sediments may occur only in samples with a weight <1 g or when AVS ≤1.5 μmol/L. Otherwise, the use of an N_2 atmosphere

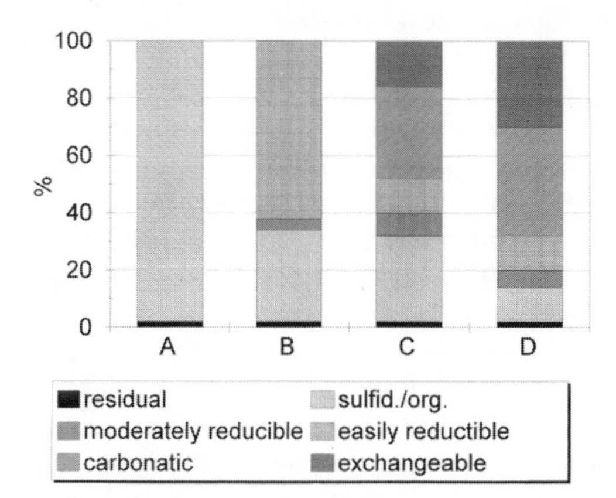

Figure 1-4 Partitioning of Cd in anoxic mud from Hamburg Harbor related to different pretreatment: (A) oxygen-free, (B) treated with U.S. EPA/Corps of Engineers elutriate test, (C) freeze-dried, and (D) 60°C oven-dried. (After Kersten and Förstner, 1987.)

does not appear to be necessary in the extraction procedures. Kersten and Förstner (1987) demonstrated that sample storage, handling, and extraction under an N_2 atmosphere by using a glove box and deoxygenated reagents will, at least, minimize changes in the extractability of metals from anoxic sediments.

1.3.6.3 Selectivity

A major drawback of chemical extractions is that they are not as selective as sometimes stated. The lack of selectivity is explained, apart from the occurrence of readsorption (especially during H_2O_2 and NH_4OAc extraction), by (1) transformation of labile phases during sample preparation, (2) variable extraction times and weight/volume ratio, (3) chelating effects and formation of soluble metal complexes, (4) soluble base metal oxides formed at initial high pH values (e.g., up to pH 13 by using NaOH or $Na_4P_2O_7$), (5) pH effects and reprecipitation of metals under oxic conditions during carbonate-bound metal extraction, (6) varying component solubility, (7) side reactions such as oxalate produced during H_2O_2 oxidation, and (8) sample contamination, losses of metals by reactions with container walls, extractant decomposition, S^{2-} formation, and AAS burner clogging, particularly during dithionite/citrate extraction. In addition, the H_2O_2 step does not discriminate between labile (i.e., metal–humate and –fulvate complexes) and stable organic phases (i.e., metals incorporated in the organic detritus) and between organic and sulfidic phases (i.e., certain sulfides and organic substances are distributed between several fractions) (Förstner, 1985; Pickering, 1986; Hirner, 1992).

Inefficiency of extraction has also been found due to (1) formation of coatings (i.e., some components do not dissolve completely, as there are often mixed coatings present in natural sediments and no clearly defined layers), (2) improper reagent use (e.g., although H_2O_2/HNO_3 dissolves Mn(IV) oxides best, it is recommended for organic matter extraction), or (3) false interpretation (e.g., metal sorbed by humic acids would have been expressed as being chelated, although it was displaced by cations) (Pickering, 1986).

Rather than specific phases, operationally defined groups of compounds, or a mixture of sedimentary fractions, with various binding forms for individual elements are extracted. Operationally defined means that the amount of metals in the final extract is considered to be equal to the concentration of metals associated with the solid component attacked by the extractant. However, repeated treatments of the sequential extraction often give a further release of metals, particularly from reducible fractions. Extractants release metals simultaneously from chemically different artificial substrates. This explains the lack of correlation which is frequently observed between a particular fraction and the compound concentration (Jenne and Luoma, 1977; Rendell et al., 1980; Surija and Branica, 1995). Further, high contents of exchangeable Mn and Fe in the NH_4OAc extract are an artifact, as the extractant also dissolves Mn oxides (Kersten and Förstner, 1987). Further,

NH_2OH–HCl also extracts part of amorphous Fe oxides, similar to boiling oxalate or DCB, which extracts both Fe and Mn oxides (Bruemmer et al., 1986). Labile sulfides may gradually dissolve when anoxic sediments are extracted by these acidic reagents. For this reason, the concentration of metals extracted by acidic NH_2OH–HCl may be an overestimate and, on the other hand, the concentration of sulfidic metals an underestimate. Furthermore, in one example it was shown that none of the trace metals spiked to a contaminated model sediment was extractable by the Tessier method in the respective fraction, which may strongly affect the results. However, if extractions now perform poorly on model systems, their performance in using real sediments is questionable.

Nonselectivity of chemical extractions is also due to the absence of thermodynamically stable, distinct solid–metal compounds in the sediments. Carbonate and easily reducible forms may be well defined, while exchangeable and organic phases are very ill defined by chemical extraction (Salomons and Förstner, 1980). Due to great heterogeneity and the different types and strengths of existing bonding sites, the recovery of metals by extractions is very reagent dependent. Both the released quantity of metals and the bonding mode between surface ions and the substrate vary with extractant composition (Pickering, 1986). A gradient of strength of associations between metals and solid particles is defined rather than the actual speciation. Experiments with individual binding substrates, different spikes, and L/S ratios resulted in a different percentage distribution among fractions, particularly for exchangeable and carbonate-associated metals, and may confirm the interdependence of extraction efficiency and the chemical behavior of the element (Kong and Liu, 1995). In addition, competitive adsorption effects may limit the selectivity of extractants (Rendell et al., 1980).

Grain-size effects may cause low extraction efficiencies, due to coatings attached to substrates of coarser grain-size classes, which are considered to be environmentally inactive and therefore sometimes neglected. Normalization of the concentration of metals with regard to conservative elements, such as Al and Sc, or the use of fine grain-size classes (i.e., <63 μm) may correct for this (Förstner, 1985).

1.3.6.4 Redistribution of metals during extraction

Readsorption is a combination of processes such as complexation, precipitation, and co-precipitation/adsorption of extracted metals. It has been reported to occur after extraction with dilute HCl (pH >1.5), 0.1 *M* NH_2OH–HCl, 0.1 *M* Na citrate (pH 4.6), 1 *M* NH_4OAc, 10% Na citrate/Na dithionate, and H_2O_2 digestion, but not with EDTA digestion (Rendell et al., 1980; Morrison et al., 1989). Readsorption effects mean that the extracted concentration of metals is less than the predicted. For example, Cu spiked onto unpolluted artificial sediments was released by 1 *M* $MgCl_2$ but readily readsorbed by the many active sites of other substrates and, therefore, extracted later again in residual phases (i.e., clays) (Martin et al., 1987). Easily reducible forms of metals released from hydrous Mn oxides by NH_2OH–HCl extraction were relatively low, due to readsorption onto hydrous Fe oxides (Elsokkary et al., 1995). In addition, organic matter may readsorb metals after the dissolution of hydrous oxides (Martin et al., 1996). Losses of metals have been observed during Na citrate/Na dithionate extraction due to the formation of free S^{2-} and metal sulfide precipitates onto container walls. Thus, interpretation of results of the sequential extraction procedure must take into account that adsorption of metals released by weak extractants may erroneously control the metal concentration in stronger extractant solutions (Rendell et al., 1980).

Increasing solid/solution ratios results in more available binding sites and hence an increased adsorption/chelation, even at pH <5 in the presence of humic acids, and also due to increasing pH during extraction, as acid consumption increases. It seems that readsorption is primarily a function of the final pH (Trefry and Metz, 1984). At low pH, organic substances may act as sorbents, while at high pH, the release of organic metal complexes may occur (Figure 1-5). Recrystallization occurs, for example, due to low acidification, which dissolves FeS but produces insoluble metal sulfides and humic precipitates (Martin et al., 1996). For this reason, the pH of extraction solutions should be kept constant, which is, however, difficult, particularly for carbonate-rich sediments. The variations in pH can be partly overcome by the use of buffering solutions, which, at the same time, may reduce readsorption due to their complexing ability (Pickering, 1986; Obermann and Cremer, 1992).

Readsorption will enhance with decreasing dissolved metal concentration and extraction time and will depend strongly on the strength of extractant. Further, matrix effects (i.e., the relative abundance of different sediment components) may limit extraction efficiency (Martin et al., 1987). It is considered that readsorption of metals may be reduced in contaminated

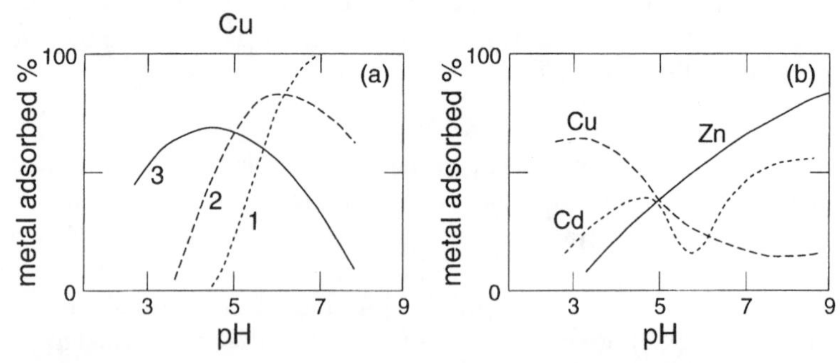

Figure 1-5 Adsorption of trace metals on alumina (left) for Cu in the presence of organic carbon (1 = none, 2 = 9.4 mg/L, 3 = 46.8 mg/L) and (right) for Cu, Zn, and Cd with fulvic acid (80 mg organics per liter). (After Davis, 1984; Bourg and Schindler, 1985.)

sediments, where many metal binding sites are already occupied (Rendell et al., 1980).

Careless use and interpretation of the extraction data without appreciating their limitations will generate erroneous and misleading information. Unfortunately, the ability of extraction techniques to simulate the natural processes is contrary to the use of strong (fast reactions) and weak (slow reactions) reagents (Martin et al., 1987).

1.4 Discussion and recommendations

Despite obvious limitations, chemical extraction procedures can be used to (1) establish solid phases, such as oxides and carbonates, which bind metals; (2) estimate the bioavailability of metals; (3) estimate release and binding processes; (4) compare the behavior of different solids; (5) evaluate pollution levels; (6) assess the mobility of contaminants; (7) determine correlations between biological uptake and sediment composition; (8) describe leaching effects; and (9) assist in geochemical prospecting (Pickering, 1986). At present, it is known that the majority of trace metals do not occur as discrete phases, but are distributed between bulk components formed by the more abundant elements, such as Fe, Mn, Ca, S, and organic C (Burton and Statham, 1982).

The major reasons for metal precipitation and complexation reactions in sediments are (1) oxidation of reduced species, (2) reduction of metals by organic matter, (3) SO_4 reduction, (4) alkaline-type reactions, (5) adsorption/co-precipitation, and (6) ion exchange (Novotny, 1995). The formation of definite metal compounds requires (1) a high concentration of dissolved metal, (2) low solubility of the compound to be formed, (3) neutral to alkaline pH, (4) low content of specific sorption sites, and (5) fewer substances that may prevent precipitation. However, this is rarely found in aquatic systems, because metals may be adsorbed before the formation of definite compounds takes place (Rose et al., 1979).

1.4.1 Main binding forms and mechanisms

Master variables, such as pH, redox potential, and organic matter, will determine the mode of metal binding to sediment particles. This type of binding will subsequently affect metal mobility, reactivity, and bioavailability.

1.4.1.1 Sorption

The sorption of trace metals in sediments is based on the following main binding mechanisms (Jenne and Luoma, 1977; Salomons and Förstner, 1980):

1. Physical sorption by relatively weak van der Waals forces (about 1 kcal/mol) and ion–dipole and dipole–dipole interactions near the surface and by isomorphic substitution of cations in surface atomic

layers and pores, resulting in three-dimensional coordinative complexes of the adsorbed metal ion
2. Chemical sorption by chemical association (diffusion exchange, coprecipitation) of ions or molecules with surface ligands of Fe or Mn oxyhydrates, carbonates, sulfides, and phosphates (e.g., Ag diffuses into Mn oxides and substitutes/exchanges with structural Mn, K, Na) or by hydrolytic adsorption [e.g., $Si{-}OH + Fe(OH)_3 = Si{-}O{-}Fe(OH)_2 + H_2O$)]
3. Ion exchange by the compensation of negative or positive charges in the mineral lattice [e.g., those of Si(OH), $Al(OH)_2$, Al(OH) groups in clays; FeOH groups in Fe hydroxides; or carboxyl and phenolic groups in organic substances] with exchangeable cations of opposite charge, with protonation not only at surfaces but also at internal surfaces/pores
4. Precipitation of dissolved compounds
5. Complexation with dissolved and solid organic matter
6. Fixation in inert lattice positions of minerals

The relative importance of these processes is outlined in Table 1-7.

It is interesting to note that almost 100% adsorption occurs within 1 to 2 pH units (i.e., adsorption edge) depending on pH and adsorbent and adsorbate concentration (Figure 1-6). Translated to natural systems, this may suggest that already small pH variations (e.g., due to acidification of the ecosystem) may cause a sharp increase or decrease of the dissolved metal fraction (Reuther et al., 1981; Förstner, 1985). However, the distribution of adsorbed metals among operational fractions is due not only to various binding strengths but also to kinetic effects. For example, desorption

Table 1-7 Relative Importance of Main Sorption Processes and Substrates in the Sediment

Sorption process	Hydrous Fe and Mn oxides	Reactive organic matter	Carbonate	Metal precipitates	Detrital mineral, organic residues
Physicosorption	X	X			C
Chemosorption	XXX	XX	XX	X	CC
Ion exchange	XXX	XX			CC
Precipitation				XXX	
Complexation	CC	XXX	C		CC
Mineral fixation	X				XX

Note: C, CC = effective via coatings.

After Salomons and Förstner, 1980.

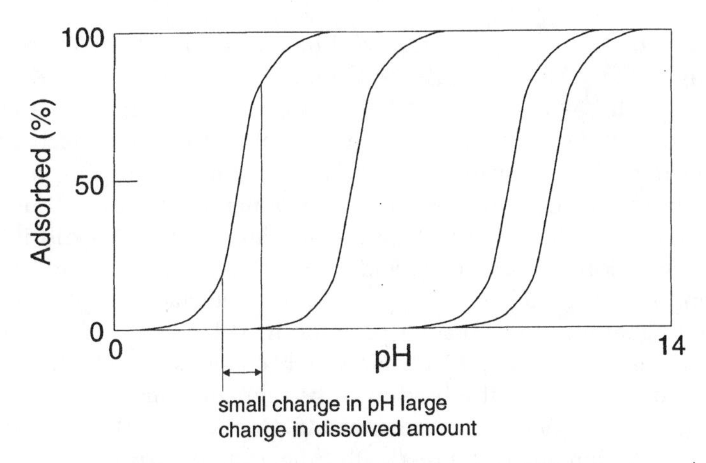

Figure 1-6 Adsorption of metals on solid surfaces, with different adsorption edges at different pH values. (After Salomons, 1995.)

may need more time to reach equilibrium than adsorption (i.e., both during extraction and naturally occurring) (Martin et al., 1996).

Different binding mechanisms may exist for alkali and heavy metals. Generally, binding constants (k_M) decrease in the order Pb, Cu, Cd, Zn > Ca > Mg >> Na, NH_4 (Pickering, 1986). However, these constants are not a unique factor for a given metal substrate, but will decrease with gradual filling of the binding site. This is explained by the occurrence of low- and high-energy sorption sites at mineral surfaces. The latter are fewer in number and become limited first (Förstner, 1987). This observation is supported by the concept of specific and nonspecific adsorption. Nonspecific adsorption is identical to ion exchange, while specific adsorption also includes no-charge-dependent, stable bondings onto solid surfaces (e.g., by complexation reactions, i.e., "surface precipitates"). In contrast to electrostatic adsorption onto charged surfaces, which primarily depends on the solute concentration and available exchange sites, specific adsorption occurs according to thermodynamic laws (Obermann and Cremer, 1992).

Among the various theories to describe the adsorption of metal ions on solid surfaces, the surface complexation model, in which oxide surfaces act as weak acids or bases in solution undergoing protonation due to pH change (Stumm and Morgan, 1981; Davis and Kent, 1990), is favored today. The oxide surface can react with other ions in solution (i.e., complexation), thereby releasing protons. Specific adsorption can be seen as a classical solution reaction with surface sites instead of dissolved ligands, according to Me + surface H = surface Me + H (Bourg, 1995). For example, Zn associated with specific sites was bound to surface OH groups of Fe–Mn–Al oxides and not available by cation exchange. Anions such as AsO_4 or Cr_2O_4 displace OH and OH_2 ligands on Fe and Al oxide surfaces by complex binding. Specific adsorption increases with the ability of metals to form hydroxy complexes according to Cd < Ni < Co < Zn << Cu < Pb << Hg. Thus the pK value (of the reaction $Me^{2+} + H_2O = MeOH^+ + H^+$) may characterize the adsorption behavior of different metals (Bruemmer et al., 1986). This mechanism becomes even more complicated when considering that adsorption also occurs within surface layers or in the interior of minerals by slow diffusion (i.e., electric double-layer theory). As an example, a three-step diffusion-controlled adsorption has been observed for goethite, as well as for Mn oxides, illites, and smectites, in the following order: (1) surface adsorption, (2) diffusion into goethite, and (3) adsorption on inert lattice positions inside the mineral (Bruemmer et al., 1986). However, changes in pH–Eh may bring about the dissolution of the binding substrate and thus mobilize complex and stable bound metals (Obermann and Cremer, 1992). Reversibility of adsorption phenomena decreases with time, due to hydrolysis and specific adsorption, as demonstrated by the extractability of natural stable Mn in sediments of the lower Rhone River in France, which was lower than that of artificial ^{54}Mn supplied from radioactive emissions of nuclear power and reprocessing plants (Förstner and Schoer, 1984).

Different types of adsorption curves can be observed for trace metals. In general, they may all follow a Langmuir isotherm, that is, with a definite adsorption maximum (e.g., Co in Figure 1-7). However, experimentally derived adsorption isotherms cannot be extrapolated to field conditions, where competition reactions, dissolved complexation, and hydrolysis occur. For this reason, the isotherm approach does not seem very practical to predict the adsorption behavior of metals in natural systems (Bourg, 1995).

1.4.1.2 Hydrous oxides

Hydrous Fe and Mn oxides occur mainly as amorphous, microcrystalline, or "aged" crystalline oxyhydroxide coatings on solid particles and, to a lesser

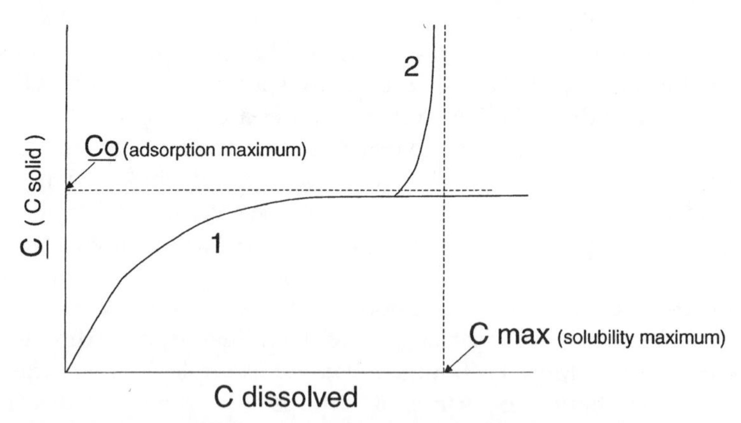

Figure 1-7 Sorption isotherm and maximum solubility: 1 = adsorption curve of the Langmuir type and 2 = precipitation curve. (After Bourg, 1995.)

extent, as discrete compounds in sediments (Förstner, 1985). As coatings, they are a prime medium, other than organic matter, for the sorption of dissolved both inorganic and organic constituents. Due to their high surface site density and area (up to 300 m^2/g), hydrous oxides, together with organic matter (up to 1900 m^2/g), are considered to be highly reactive adsorbents (e.g., surface site area for carbonates is only <1 m^2/g) (Surija and Branica, 1995). Moreover, they may completely change the behavior and reactivity of other solids, such as clays, carbonates, quartz, or feldspars (Förstner, 1987). Metal sorption on oxides has been described as surface complexation (discussed above). To interpret the adsorption of metal ions on oxidic surfaces, the empirical hydrolysis–adsorption model after James and Healy (1972) was considered. The model suggests that strong adsorption occurs if log β1 is high (i.e., the hydrolysis constant for the first OH group attached to the metal ion, reflecting the degree of hydrolysis). Different degrees of hydrolysis may explain varying reversibilities observed for sorption processes (for example, low desorption may indicate low reversibility when pH decreases due to hydrolytic adsorption).

1.4.1.3 Organic matter

There is a high content of weak acid carboxylic and phenolic functional groups in the sediment organic matter available to bind metals. These surface-active groups (i.e., COOH, C=O, NH_2, SH_2) are provided by (1) living organisms, (2) dead plant and animal material (i.e., detritus), and (3) low-molecular-weight organics, which occur as coatings on clay and metal oxides (Table 1-8).

Negatively charged surface coatings of organic matter may interact with metals by different binding mechanisms, such as ion exchange, protonation, van der Waals forces, coordination complexation, and hydrophobic bonding (Förstner, 1987). Similar to Fe and Mn oxides, these coatings may completely change the adsorption behavior of metals and inorganic substrates. They are interrelated with inorganic Fe and Mn hydroxide coatings (Salomons and Förstner, 1984). Jonasson (1977) established an *order of binding strength* be-

Table 1-8 Main Forms of Organic Compounds in the Sediment

Fraction	Components
Humic material	Dissolved or colloidal humic–fulvic acids
Organic coatings	Precipitates/complexes of humic material on mineral particles
Living organisms	Bacteria, fungi, benthos
Biotic detritus	Dead organic material
Nonhumic substances	Carbohydrates, proteins, peptides, amino acids, fats, waxes, resins

tween metals and humic/fulvic acids in sediments (Hg > Cu > Pb > Zn > Ni > Co) and in pore water (Cu > Pb > Cd). A similar sequence was found for metals bound to humic and planktonic material (Cu > Zn > Ni > Co > Mn and Pb > Cu > Cd > Zn, respectively) (Rashid, 1974; Förstner et al., 1983). For example, the H_2O_2 oxidizable metal fraction extracted from silty clay from Hamburg Harbor increased in the order Mn < Fe < Cr = Ni < Zn < Pb < Cd < Cu, corresponding well with the Irving–Williams sequence for complex stability, but also with the stability of metal sulfides (Förstner and Kersten, 1988). With regard to the type of binding, it was suggested that about one-third of metals bound to organics are exchangeable and two-thirds in a complexed form, with binding capacities of 200 to 600 meq metals per 100 g humic substance (Rashid, 1971). Fulvic acids are known to possess two to six times greater chelating capacity than humic acids (>600 dalton). However, the complexing and sorption properties of dissolved and solid organic matter in the sediment vary together with the solid/solution equilibria of metals (i.e., similar to AVS as described below), due to diurnal and seasonal changes in biological activity (Förstner and Kersten, 1988).

Similar to adsorption of metals on inorganic surfaces, organic complexation is very pH dependent. For example, complexation of Cu by humic acids reaches a maximum at pH 2.5 to 3.5, in contrast to fulvic acids with an optimum at about pH 6 (Figure 1-5). In general, solid organic substances are stronger adsorbents at low pH than inorganic substrates, and ligands, sorbed to surfaces, may generally increase the sorption capacity of sediments (Herms and Bruemmer, 1980).

Often the acidity constant (pk) of organic-rich sediments is similar to that of humic acids, suggesting that a great deal of the proton-binding capacity (i.e., about 60%) of sediments can be ascribed to humic acids. If these ligands, as well as EDTA and DTPA, are not adsorbed, they may bind dissolved metals, thereby decreasing their adsorption capacity. However, dissolved inorganic and organic ligands complex not only dissolved but also solid cations from the particle surface, therefore competing with solid components for binding the metals (Förstner, 1985).

In anaerobic sediments, organic substances are electron donors. For example, Cr(VI) reduction increases with the organic matter, and reduction of Cr(III) is mainly hampered by the presence of organic material (Losi et al., 1994).

1.4.1.4 Acid-volatile sulfide

AVS is considered to be a key binding phase for metals in anaerobic marine and freshwater sediments, controlling both pore water concentration and the bioavailability of metals to benthic organisms. The formation of AVS appears to be strongly related to the amount of organic matter and to sulfate in the sediment (e.g., microbial oxidation of organic matter supplies the necessary electron for SO_4^{2-} as the terminal electron acceptor). This latter mechanism may help to explain the following observations in natural sediments: (1) low AVS levels occur during winter, probably due to low biomass production; (2) sulfate content in pore water limits the AVS; (3) reoxidation of Fe^{2+}; and (4) pyrite formation reduces AVS (Allen, 1993; Leonard et al., 1995).

High concentration of metals in the pore water often correlates with a low AVS content in the sediment (Brumbaugh et al., 1994). AVSs involve amorphous FeS and MnS, mackinawite (FeS), greigite (Fe_3S_4) (a precursor to FeS_2), and pyrrhotite and represent the total reactive pool of solid-phase sulfides available to bind metals with a lower solubility (Zhuang et al., 1994). Titration experiments showed that metal sulfides are less soluble than iron monosulfides (for example, Zn, Cu, and Pb are scavenged until the AVS in the sediment is saturated) (Casas and Crecelius, 1994). When the concentration of S^{2-} in the pore water becomes depleted (e.g., due to binding to other metals), FeS dissolves. Divalent metals displace Fe^{2+} in the FeS and form more insoluble MeS and soluble Fe^{2+}. One metal can exchange with another, if the sulfide formed is more insoluble (e.g., $Cu^{2+} + ZnS = CuS + Zn^{2+}$). This process occurs quite rapidly (within minutes to hours) and involves any metal less soluble than FeS. It is suggested that saturation of the binding capacity is indicated at an SEM/AVS ratio >1, which may cause a flux of metals into the pore water.

However, the sediment-binding capacity for metals can be greater than indicated by AVS, also suggesting the presence of other important binding sites. This was shown for Cu, in contrast to Zn and Pb, where AVS was the strongest contributing binding substrate (Casas and Crecelius, 1994). The importance of other binding phases was also demonstrated for spiked Cd, which remained mostly partitioned in the sediment, although molar concentrations yielded Cd/AVS >1 (Zhuang et al., 1994). In particular, in contaminated sediments, binding phases other than AVS must be considered (e.g., Fe/Mn oxides, organic matter, etc.). If AVS is present in a high concentration and metal recovery is low, as observed for Cu and As, the SEM/AVS concept appears questionable, at least for the above metals (Brumbaugh et al., 1994). In fact, no direct correlation has been found by Kemble et al. (1994) between AVS and the absolute amount of SEM in the sediment. In addition to AVS, other forms of labile sulfides, such as CRS, exist (Table 1-5). Three different forms of volatile sulfides with complex reaction equilibria have been measured in sediments: molecular H_2S, AVS, and heat-volatile sulfides (Brouwer and Murphy, 1995).

Sediment aeration, for example, by resuspension, bioturbation, or dredging, rapidly oxidizes AVS within 1 day, resulting in an increase in Fe and Mn oxides. The concentration of AVS becomes nondetectable after several days, and large amounts of acid are produced, particularly in the case of low carbonate concentration in the sediment. It was demonstrated that AVS-bound Cd was rapidly released, while Cd bound to pyrite dissolved only after long periods of aeration (Zhuang et al., 1994).

1.4.2 Mobility and reactivity of metals

Sediments represent a substrate with a complex network of heterogeneous solid–solution reactions. The solubility of metals depends on the pore water composition, the geochemical phases of sediment to which metals are bound, and the available sites for binding the metals. Metals can be released into the dissolved phase and subsequently readsorbed. The release is equilibrated by the rate of release and scavenging of the metals. The increased anthropogenic input of trace metals into aquatic sediments has favored the formation of thermodynamically less stable metal phases. This is confirmed by studies with metal spikes or artificial isotopes, such as ^{54}Mn and ^{137}Cs, which show that newly introduced metals form more labile bindings with the sediment substrate than their naturally occurring metal equivalents (Förstner and Schoer, 1984). Processes that occur during early diagenesis, such as transformation of organic matter, O_2 consumption, NO_3/SO_4 reduction, aging of minerals, and the establishment of new solid/solution equilibria, gradually cause a stabilization of the bonding between the metal and substrate (Förstner, 1987). Residual metals are considered to remain unavailable for at least more than a year to a decade (Jenne, 1977).

The following factors may influence the mobility of metals in sediments (Pickering, 1986):

1. *Salinity*: Salts of alkali or earth–alkali elements may displace loosely bound metals or form soluble metal–salt complexes.
2. *Acidity*: H^+ on surfaces or ligands with a high proton affinity, such as functional groups like OH, SH, and COOH, may compete with metals for exchange and sorption sites or partially dissolve $CaCO_3$ and other metal precipitates (e.g., acid rain, acid mine drainage, NH_4^+ fertilizer).
3. *Redox potential*: Bacterial activity or eutrophication reduces the concentration of oxygen, resulting in the reduction of Fe(III) and Mn(IV) with subsequent dissolution of Fe and Mn oxides and the release of metals.
4. *Complex formation*: Natural (e.g., hydroxy, fulvic, and carboxylic acids) and synthetic complexing agents (e.g., EDTA, NTA, DTPA) convert sorbed metals into soluble metal complexes.
5. *Biochemical transformations*: Metals are translocated from solid abiotic to biological systems (e.g., by alkylation), with a transfer within food chains (e.g., methylmercury).

Empirically, it is considered that metal solubility is increasing in systems either with a high sulfur content, low pH, and moderate to high redox potential or with a low sulfur and low redox potential. In contrast, a high pH (i.e., low H^+ concentration) and oxygen content minimize the solubility of Fe and Mn (Förstner, 1995). In addition, metal anions and organically

Table 1-9 Classification Scheme for the Mobility of Metals in Relation to Their Extractability

Binding forms	Mobility
Exchangeable	Very easy mobility
Carbonatic	Easily mobilizable
Amorphous crystals	Less easily mobilizable
Crystalline oxides	Moderately mobilizable
Sulfidic	Heavily mobilizable
Mineral lattice	Very heavily mobilizable

After Brasser et al., 1995.

complexed cations may keep metals in the solid state under moderately acidic conditions (Bourg and Loch, 1995). A simple classification of the mobility of metals with regard to their extractability is presented in Table 1-9.

The capacity of sediments to adsorb or release metals depends strongly on their particular composition. For this reason, characterization of the properties controlling the capacity, such as buffering capacity, is indispensable for understanding and prediction of mid- and long-term mobilization of metals (Salomons, 1995).

1.4.2.1 Influence of pH changes

Generally, metals in sediments may become mobilized with the decrease of pH. However, many trace elements are amphoteric in nature (i.e., they dissolve at low and high pH) and are, by and large, insoluble at neutral pH (i.e., pH 8) (Eighmy et al., 1995). Significant metal mobilization occurs for most metals at pH <5, in contrast to most natural organics, which generally become mobilized at pH >7. At low pH, humic substances are mainly not acid soluble and immobile (Figure 1-5). At high pH values, the dissociation of acid groups leads to increased polarity and solubility, while mobilization of metals may occur at these pH values (Obermann and Cremer, 1992).

Therefore, the ANC (i.e., the $CaCO_3$ content) appears critical to the fate of metals in sediments. Acid consumption curves will reflect the slow, long-term release of metals from the sediment (Förstner, 1995). Oxidation reactions, such as oxidation of NH_4^+, FeS, or FeS_2 and organic matter, are a primary source of protons in the sediment. Oxidation of organic matter will increase the pCO_2 and, subsequently, the solubility of $CaCO_3$. Further, it may produce organic acids, resulting in an increasing consumption of the ANC (Johnson et al., 1995).

The buffering capacity of the sediment is strongly affected by redox reactions. For example, the reduction of sulfate increases the solid APP and decreases the ANC due to a decrease of HCO_3^-. The APP is made up by the oxidation of sulfides and organic matter, whereby 1% FeS_2 in sediment may generate as much protons as sediment with 5% organic carbon content (Swift, 1977). This example illustrates the interrelationship between dissolved and solid speciation in the sediment (i.e., increase of solid APP causes a decrease of aqueous ANC).

1.4.2.2 *Influence of redox changes*

In reduced systems, nonlinear, delayed processes dominate and may restrict prognosis of long-term mobilities of metals. However, a predictable sequence of e^- acceptors, such as O^{2-}, NO_3^-, metal oxides, SO_4^{2-}, S^{2-}, and CH_4, occurs at the surface of water bodies during the microbial degradation of organic matter. This determines the redox potential and may alter pH (Meyer et al., 1994). In addition, organic carbon is also oxidized by the reduction of MnO_2 in the sediment (Froelich et al., 1979). Decomposition of organic matter will consume O_2 and produce CO_2 and H_2CO_3 and hence increase the dissolution of $CaCO_3$. In anoxic environments, metal solubility is controlled by the presence of S^{2-}. However, if the concentrations of S^{2-} and H_2S are low, carbonates or phosphates may control the solubility of metals.

Changes from reducing to oxidizing conditions that involve sulfide transformation and more acidity increase the mobility of "chalcophilic" elements, such as Hg, Zn, Pb, Cu, and Cd, in contrast to Mn and Fe, which become oxidized, hydrolyzed, and precipitated. The mobility of Fe and Mn increases with reducing conditions (e.g., due to the formation of Fe^{2+} and Mn^{2+}). This is also true for acidification effects, although metal anions may be more mobile under neutral–alkaline conditions (Plant and Raiswell, 1983). During FeS oxidation, a part of Fe^{3+} becomes hydrolyzed, forming $Fe(OH)_3$, or is complexed by organic matter. Contrary to this, organic matter is considered to inhibit the oxidation of Fe^{2+} (Obermann and Cremer, 1992).

In most aquatic sediments, Fe oxyhydroxides, and to a lesser degree Mn oxides, and reducible organic matter contribute as electron acceptors to the OXC. For example, in marine systems, there is a diffusion of sulfate from the water down to the bottom sediment, where all Fe is transformed to FeS. In freshwater systems, however, less sulfate is available, and therefore only part of the Fe^{3+} is transformed to FeS. Most of the available reduced Fe(II) is converted into $FeCO_3$ and may act as a buffer against a pH decrease. Such suboxic conditions are characterized by the formation of $FeCO_3$ minerals (Bourg and Loch, 1995). Further, electron transfers, such as the reduction of sulfate to sulfide according to $S^{2-} + 2H^+ = H_2S$, can be considered acid-consuming/producing processes. In reduced systems, part of the exchangeable Fe^{2+} may also displace Ca and Mg. After aeration, Fe^{2+} is oxidized and H^+ is produced, removing Ca and Mg (i.e., ferrolysis) (Brinkmann, 1970).

It is suggested that the reaction kinetics for the initial mobilization of metals from solids is largely controlled by microbial activity (Salomons, 1995). Many solid–solution reactions, such as redox reactions, are kinetically controlled and therefore are slow processes. However, it is suggested that the reaction kinetics for the initial mobilization of metals from solids is largely controlled by microbial activity (Salomons, 1995). Microorganisms, such as nonphotosynthetic bacteria, use reduction processes, such as the decomposition of organic matter, as a source of energy for their own metabolism and may catalyze the reaction. For example, *Thiobacillus ferrooxidans* increases reaction rates by a factor of up to 10^6. In addition, many redox processes are accelerated after adsorption of the reactive metal species. For example, almost instant reduction of Cr(VI) was observed after adsorption to organic matter and ferrous iron in soils and sediments, but reoxidation of Cr(III) by Mn(IV) was also reported (Losi et al., 1994). In contrast to redox reactions, acid-based and adsorption reactions are relatively fast (Obermann and Cremer, 1992). Early diagenetic processes may enhance metal mobility at the beginning, due to the mineralization of organic matter with the subsequent drop in pH and production of HCO_3^-. However, diagenetic aging of minerals counteracts these processes by condensation of hydroxides, recrystallization, and migration of metal ions into mineral lattice positions, which may lead to a stronger fixation of metals in the sediment (Salomons, 1980).

1.4.3 *Bioavailability and toxicity*

The ultimate goal of the many attempts to determine metal speciation is to relate its results to the bioavailability of metals. However, there is no simple relationship between speciation and bioavailability of metals in sediments. While the bioavailability of waterborne metals is directly related to existing species, this is not possible for sediment-associated metals. Uptake of metals from solids by biota is slower than that from solutes. However, the pH of the digestive tract of organisms living on the surface or within the sediment is considered to be critical for the final uptake of particulate metals (e.g., the pH of the digestive tract of invertebrates is 5 to 8 and that of vertebrates is approximately 3) (Pickering, 1986). In fact, there are only a few studies reporting a direct correlation between total metal concentrations in sediment and in biota. For example, total concentration of Ag, Cd, Co, and Zn correlated significantly with concentrations found in the macrophyte *Fucus vesiculosus*, and there was a correlation of Cu, Co, and Pb in the sediment and in the polychaete *Nereis diversicolor*. However, no correlation was found between the concentration of Cu in the sediment and in the clam *Scrobicularia plana* (Luoma and Bryan, 1982).

1.4.3.1 *Leachability of metals*

Bioavailability can first be estimated by the leaching of metals, which may simulate different uptake modes. However, there is no quantitative link between the extractability and bioavailability of a metal. "Leachable" metals may not be bioavailable, as the specific uptake mechanisms by which organisms selectively participate in removing metals from solids are largely unknown (Luoma, 1983). As a first approximation, "available fractions" in the sediment may characterize particulates with a high surface-to-weight ratio, such as organics, hydrous oxides, clays, etc.

Many studies have been carried out on the leachability and uptake of metals by plants. Metals extractable by EDTA and DTPA are considered to be an estimate of the plant-available metal pool (Bruemmer et al., 1986). Significant correlations were found between the uptake by plants and extracting agents. These include extractions by water, 1 *M* Ba (or Mn) Cl_2, 1 *M* NH_4NO_3, HOAc, 1 *M* NH_4OAc, 1 *M* CH_3COONH_4, acid $(NH_4)_2C_2O_4$, 0.1 *M* H_3PO_4, 0.1 *M* HCl, and DTPA (Pickering, 1986). The use of $CdCl_2$ is promising to predict the plant-available portion of total Zn and Cd (Sauerbeck and Styperek, 1985).

Particular acid or chelating extractants can simulate the release and uptake of metals from particles in the digestive tract of animals (Morrison et al., 1989). Luoma and Bryan (1982) suggested that 1 *M* HCl sediment extracts would provide sufficient information on the bioavailability of metals, such as Ag, Cu, Co, and Cd, and less for Zn (Krantzberg, 1994). Luoma and Bryan (1978, 1979) found a good correlation between the uptake of Pb by the clam *S. plana* and the Pb/Fe ratio in 1 *M* HCl sediment extracts. Tessier et al. (1984) obtained a significant correlation between Pb in *S. plana* and the Pb/Fe ratio for Pb in the "exchangeable + carbonate + Fe/Mn oxide" fractions and for oxide-associated Fe. Gunn et al. (1989) established a significant correlation between Pb in the soft tissue of tubificid worms and exchangeable Pb in the sediment. The concentrations of metals in fish were significantly correlated with those extracted by NH_2OH–HCl, with the exception of Al (Krantzberg, 1994). The concentration of Zn in the clam *Macoma baltica* was controlled by a competitive partitioning of this metal between NH_2OH–HCl-extractable Fe oxides and NH_4OAC-extractable Mn oxides. The concentrations of Ag and Co in clams and tubificid worms correlated best with the HCl-extractable fraction, but not with total concentrations or with any other metal phase in the sediment (Luoma and Bryan, 1982). These examples indicate the high competition for metal uptake that exists between organisms and the sediment substrate. Selective uptake mechanisms of the biota compete with active surface sites of the sediment substrate (Reuther, 1992).

1.4.3.2 Controlling factors

The type and composition of the sediment substrate are considered to affect metal availability. Uptake by biota depends on the strength of bondings between a metal and the sediment, which may vary for different substrates and metals and with master variables, such as redox potential and pH. Solid organic matter has a greater binding capacity (i.e., reduced availability and toxicity) at low pH than mineral substances (Bruemmer et al., 1986). For example, a negative correlation was established between total organic carbon and 1 *M* HCl-extractable Fe in the sediment, which was reflected by a low metal bioavailability. In contrast, an inverse correlation was reported to exist between the carbonate content (i.e., hardness) in the sediment and the uptake of Cd by animals. Organic ligands may enhance metal mobility but generally decrease its uptake by biota, with some exceptions (e.g., cobalamine) (Jenne and Luoma, 1977). In addition, varying substrate concentrations cause the observed temporal and spatial variations of the partitioning and bioavailability of metals.

AVS is used increasingly as an indicator to predict the bioavailability and toxicity of metals to biota which is in contact with anoxic sediments. Di Toro et al. (1990) even proposed AVS as a toxicity-predicting tool. Field and laboratory studies showed that an SEM/AVS ratio <1 will cause $<50\%$ mortality in test biota, in contrast to an SEM/AVS ratio >1, which may give rise to $>50\%$ mortality. In the latter case, divalent metals may become soluble and bioavailable in the pore water. Recently, the AVS concept was introduced by the U.S. EPA as a criterion, considering that nonbioavailability will exist at a surplus of sulfide ions (Di Toro et al., 1990). Indeed, both laboratory toxicity and bioaccumulation tests, as well as evaluation of the benthic community, identify sediments, mostly oxic, with a low AVS as the most impacted area regarding the health of the biota (Brumbaugh et al., 1994).

However, tests using the amphipod *Hyalella azteca* yielded no significant mortality at a Cd/AVS >1, despite a pore water toxic unit concentration (i.e., pore water concentration/LC_{50}) of >1. This has been partly explained by the formation of nonbioavailable organic/polysulfidic Cd complexes (Leonard et al., 1995). On the other hand, bioaccumulation has been observed for SEM/AVS ≤ 1, indicating that this approach may be useful to predict dissolved pore water concentrations, but may under- or overestimate the truly bioavailable fraction of the metal. One reason may be, for example, the action of burrowing organisms, which brings about oxidized microzones in the sediment with the subsequent release of metals, such as Zn and Cu. Although the concentration may be low, these metals may bioaccumulate, even at SEM/AVS ≤ 1. To elucidate the exact mechanisms, further speciation studies with different types of sediments in combination with bioassays are necessary. In any case, AVS does not control the bioavailability of metalloids, such as As or Se (Ingersoll et al., 1994).

In addition to feeding habits (i.e., bioavailability of particulate metals seems more related to detritus-feeding organisms, such as tubificid worms, than to filter-feeding organisms, such as clams), metal–metal interactions (i.e., antagonistic, synergistic, and additive effects) have to be considered. For example, a strong negative correlation was observed between the concentration of Ag in clams and the Cu concentration in the sediment (i.e., antagonistic effect). A regression analysis showed that the Ag and Cu concentrations in the sediment explained about 77% of the variation of Ag in clams (Luoma and Bryan, 1982).

Further, some particular forms of heavy metals are able to increase bioavailability by using the transport ways of nutrients with a similar chemical structure, to pass through biomembranes (for example, CrO_4/SO_4, AsO_4/PO_4, CrO_4/NO_3, Cd/Ca) (Reuther, 1986).

It is understood that the free ion concentration may best reflect the biologically available form. For this reason, pore water toxicity tests are considered to be appropriate to assess sediment quality (Carr and Chapman, 1995). Indeed, bioassays showed a significant mortality at a pore water concentration exceeding the water-only LC_{50} concentration. A more accurate toxicological measure may be the use of a pore water toxic unit concentration (i.e., pore water/LC_{50}) as indicated above, in addition to the SEM/AVS ratio (Casas and Crecelius, 1994). For example, a good correlation was documented between the pore water toxic unit and the mortality of sediment-dwelling worms (Pesch et al., 1995). However, pore water toxicity strongly depends on the particular form of the dissolved metal. For example, dissolved As(III) is considered to be about 50 times more toxic than As(V) and some hundred times more toxic than organoarsenicals, such as monomethyl arsenate and dimethyl arsenate (Morrison et al., 1989). In addition, Cr(VI) is known as a strong mutagen and is very mobile in solid/solute systems, in contrast to Cr(III), which is not a mutagen and is highly immobile and insoluble at pH >5.5 (Losi et al., 1994).

1.4.4 *Standard procedures in metal speciation*

Any approach chosen to determine the speciation of metals in sediments must relate to a defined problem. The assessment of critical metal fractions (i.e., bioavailable pools) should include pore water and solid surface analysis, single leaching (i.e., short term), and chemical extraction (i.e., short to long term), in combination with bioassays (i.e., bulk sediment toxicity tests). The selected test system should simulate relevant biological uptake systems, for example, by including acid (i.e., digestive gut) or chelating agents (i.e., cell surfaces) or synthetic membranes.

With regard to chemical leaching, it is recommended to (1) design the extraction sequence by improving the "selectivity" for a particular sediment; (2) reduce readsorption by a dynamic extraction process, for example, in a flow cell or by ion-exchange systems; (3) include washing solutions; and (4) repeat extractions due to mutual inclusion between inorganic and organic phases (Hirner, 1992).

Ideally, the leaching reagent should extract all metals from the particular sediment phase and not affect metals associated with other phases, therefore maintaining the integrity of individual metal species (Byrdy and Caruso, 1994). The concentration and composition of the extractant should, at best, reflect or simulate particular environmental conditions with regard to pH, redox potential, and complexing capacity, which may influence metal release and fixation. For example, weak acids may imitate the effect of acid depositions (i.e., acid rain), oxidizing reagents may imitate the decomposition of organic matter or exposure to air, and reducing agents may resemble oxygen-deficient conditions (Förstner et al., 1989b).

There is a growing need to include solid speciation criteria in existing sediment quality assessments. This requires harmonization and standardization of the available methods. Such standardization has to consider the following aspects: the objective of the study, the type of sediment and elements of interest in accordance with the results expected, and the economic and environmental relevance of a specific site or project. For this, a database should be installed, which includes all appropriate methods covering a wide range of typical problems and analytical specifications. Standardized procedures should not be too complicated or time consuming, but highly informative. They should be a part of a more comprehensive approach to assess the quality of sediments, including biological testing and predictive numeric modeling (Förstner et al., 1989a).

As an example, two standardized leaching tests have been proposed in Germany in order to assess the groundwater contamination potential of soils. The test DIN 38414 S7 evaluates the acid-soluble, nonsilicate part of metals by acid digestion using *aqua regia*, providing an overview of the existing total metal pool. In contrast, the test DIN 38414 S4 uses only distilled water to extract water-soluble metals, giving an estimate of the initially available, short-term released metal fraction. These two standards could be combined with the recently developed pH_{STAT} test, which elucidates acid/base characteristics (i.e., the influence of the ANC on the leachability of metals in particulates). The pH_{STAT} test will further provide information on the long-term behavior of metals in relation to worst-case conditions.

In order to standardize the elution data, an "elution index" has been proposed based on the 1 *M* NH_4OAc-extractable portion of metals, which is considered to be remobilizable during a relatively short time. Comparison of release rates from oxic and anoxic sediments indicated that mobilization of metals may increase, particularly due to oxidation (Förstner et al., 1990).

To improve the analytical quality of methods (i.e., precision and accuracy), certified reference material is needed with a known distribution of metals among various well-characterized solid fractions. Such certified reference materials should match the variability and complexity of naturally occurring sediments with regard to master variables, such as particle size, organic matter content, pH, and redox potential (Suedel and Rodgers, 1994). The use of metal spikes as well as model sediments and compounds, in addition to a broadly applied extraction scheme, such as the Tessier method, would contribute to the improved efficiency and selectivity of available methods (Kheboian and Bauer, 1987; Pickering, 1986; Martin et al., 1996).

1.5 Recommendations for future studies

The definition of sediment quality criteria (SQC) for trace metals based solely on total concentrations will not meet the complex interactions that control the fate and mobility of the metals. "Criteria and standards must implicitly or explicitly incorporate principles of chemical speciation to provide appropriate numeric values" (Allen, 1993). However, chemical speciation alone will not describe biological implications, and effect analysis remains indiscriminate with regard to the cause and remediation of effects. There is a need for approaches which are capable of integrating both dimensions (i.e., mechanisms), which link potential paths of partitioning with the uptake of trace metals.

For short-term prediction, bioassays using pore water, elutriates, or bulk sediment samples may be sufficient. However, long-term prediction may be better guided by information obtained from sequential chemical extraction. In addition, future SQC should be based on elutriate tests, preferentially at pH 4, in relation to certain categories of mobility-controlling sediment parameters, such as the APP and acid-consuming capacity, as the internal (i.e., substrate) chemistry may have a strong impact on metal transport and fate.

Three main sediment parameters were selected for establishing SQC with regard to long-term effects of sediment-bound metals: (1) the APP, (2) acid consumption capacity, and (3) metal concentration in pore water. In addition, AVS and total organic carbon concepts appear promising for predicting metal bioavailability.

From the 1970s (Tessier et al., 1979) to the 1990s (Di Toro et al., 1990), the speciation of metals has remained a major characteristic when assessing the type and extent of metal contamination in sediments. Chemical extraction methods have evolved as a predictive tool to identify shifts in the pattern of main associations, which may indicate a weakening or strengthening of the binding of metals in the sediment.

According to data presented, there is a need for further research to obtain information on the following:

1. Kinetic effects that control adsorption/desorption equilibria
2. The extent to which desorption rates of metals are influenced by different solid surfaces
3. The potential of pore water analysis to predict chronic effects, in combination with the AVS concept
4. The influence of spatial and temporal variations of main sediment properties, such as ANC and AVS, on nonlinear, delayed processes
5. The potential of chemical extraction techniques to estimate the bioavailability of metals for distinct classes of organisms
6. Natural variations of the bioavailability of metals among biotic species and individuals

7. How to quantify the bioavailable part of metals in sediments in relation to competitive reactions between surface and biological uptake mechanisms (this may be useful to predict vulnerable ecosystems)
8. The influence of early diagenetic processes on the association of metals with different sediment phases
9. Methods for determination of the speciation of Al in sediment
10. Methods to extract both siderite ($FeCO_3$) and magnetite (Fe_3O_4)
11. Methods to separate humic/fulvic acids from carbonates in the sequential extraction procedure

References

Adams, D.D., Sediment pore water sampling, in *Handbook of Techniques for Aquatic Sediment Sampling,* Mudroch, A. and MacKnight, S.D., Eds., Lewis Publishers, Boca Raton, FL, 1994, 171.

Agemian, H. and Chau, A.S.Y., A study of different analytical extraction methods for nondetrital heavy metals in aquatic sediments, *Arch. Environ. Contam. Toxicol.,* 6, 69, 1977.

Allan, R.J., Metal contents of lake sediment cores from established mining areas: an interface of exploration and environmental geochemistry, *Geol. Surv. Can.,* 74(1/8), 43, 1974.

Allard, B., Håkanson, K., and Karlsson, S., The importance of sorption phenomena in relation to trace element speciation and mobility, in *Speciation of Metals in Water, Sediment and Soil Systems,* Landner, L., Ed., Springer-Verlag, Berlin, 1987, 99.

Allen, H.E., The significance of trace metals speciation for water, sediment and soil quality criteria and standards, in Proc. 2nd Eur. Conf. Ecotoxicol., Sloof, W. and de Kruijt, H., Eds., *Sci. Total Environ.,* 1, 23, 1993.

Allen, S.E., Grimshaw, H.M., Parkinson, J.A., and Quarmby, C., *Chemical Analysis of Ecological Materials,* Blackwell, London, 1974, 384.

APHA/AWWA/WPCF, *Standard Methods for the Examination of Water and Wastewater,* American Public Health Association, Washington, D.C., 1985.

Baudo, R., Giesy, J.P., and Muntau, H., *Sediments: Chemistry and Toxicity of In-Place Pollutants,* Lewis Publishers, Boca Raton, FL, 1990, 405.

Benjamin, M.M. and Leckie, J.O., Multiple-site adsorption of Cd, Cu, Zn and Pb on amorphous iron oxyhydrides, *J. Colloid Interface Sci.,* 79, 209, 1981.

Bernhard, M., Brinckmann, F.E., and Sadler, P.S., *The Importance of Chemical "Speciation" in Environmental Processes,* Dahlem Conf., Life Sciences Res. Rep. 33, Springer-Verlag, Berlin, 1986, 763.

Bourg, A.C.M., Speciation of heavy metals in soils and groundwater, and implications for their natural and provoked mobility, in *Heavy Metals,* Salomons, W., Förstner, U., and Mader, P., Eds., Springer-Verlag, Berlin, 1995, 19.

Bourg, A.C.M. and Loch, J.P.G., Mobilization of heavy metals as affected by pH and redox conditions, in *Biogeodynamics of Pollutants in Soils and Sediments,* Salomons, W. and Stigliani, W.M., Eds., Springer-Verlag, Berlin, 1995, 87.

Bourg, A.C.M. and Schindler, P.W., Control of trace metals in natural aquatic systems by the adsorptive properties of organic matter, in Proc. Int. Conf. Heavy Metals in the Environment, Athens, September 1985, 97.

Brasser, T., Brewitz, W., Bahadir, M., and Reichelt, C., Auslaugverhalten von schwermetallhaltigen Sonderabfällen in Untertagedeponien, *Müll Abfall*, 6, 388, 1995.

Breder, R., Optimization studies for reliable trace metal analysis in sediments by atomic absorption spectrometric methods, *Fresenius Z. Anal. Chem.*, 313, 395, 1982.

Brinkmann, R., Ferrolysis, a hydromorphic soil forming process, *Geoderma*, 3, 199, 1970.

Brouwer, H. and Murphy, T., Volatile sulfides and their toxicity in freshwater sediments, *Environ. Chem. Toxicol.*, 14, 203, 1995.

Brown, D.S. and Allison, J.D., MINTEQA1, An Equilibrium Metal Speciation Model: User's Manual, U.S. Environmental Protection Agency, Washington, D.C., 1987.

Bruemmer, G.W., Gerth, J., and Herms, U., Heavy metal species, mobility and availability in soils, *Z. Pflanzenernaehr. Bodenkde.*, 149, 382, 1986.

Brügmann, L., Metals in sediments and suspended matter of the River Elbe, *Sci. Total Environ.*, 159, 53, 1995.

Brumbaugh, W.G., Ingersoll, C.G., Kemble, N.E., May, T.W., and Zajicek, J.L., Chemical characterization of sediments and pore water from the Upper Clark Fork River and Milltown Reservoir, Montana, *Environ. Toxicol. Chem.*, 13, 1971, 1994.

Bruynesteyn, A. and Hackl, R.P., Evaluation of acid producing potential of mining waste minerals, *Mineral. Environ.*, 4, 5, 1984.

Burton, J.D. and Statham, P.J., Occurrence, distribution and chemical speciation of some minor dissolved constituents in ocean waters, in *Environmental Chemistry*, Bowen, H.J.M., Ed., Royal Society of Chemistry, London, 1982, 301.

Byrdy, F.A. and Caruso, J.A., Elemental analysis of environmental samples. Using chromatography coupled with plasma mass spectrometry, *Environ. Sci. Technol.*, 28, 528A, 1994.

Calmano, W., Ahlf, W., and Förstner, U., Study of metal sorption/desorption processes on competing sediment components with a multi-chamber device, *Environ. Geol. Water Sci.*, 11, 77, 1988.

Campbell, P.G.C. and Tessier, A., Current status of metal speciation studies, in *Metals Speciation, Separation, and Recovery*, Patterson, J.W. and Passino, R., Eds., Lewis Publishers, Boca Raton, FL, 1987, 779.

Canfield, T.J., Kemble, N.E., Brumbaugh, W.G., Dwyer, F.J., Ingersoll, C.G., and Fairchild, J.F., Use of benthic invertebrate community structure and the sediment quality triad to evaluate metal-contaminated sediment in the Upper Clark Fork River, Montana, *Environ. Toxicol. Chem.*, 13, 1990, 1994.

Carr, R.S. and Chapman, D.C., Comparison of methods for conducting marine and estuarine sediment pore water toxicity tests — extraction, storage, and handling techniques, *Arch. Environ. Contam. Toxicol.*, 28, 69, 1995.

Casas, A.M. and Crecelius, E.A., Relationship between acid volatile sulfide and the toxicity of zinc, lead and copper in marine sediments, *Environ. Toxicol. Chem.*, 13, 529, 1994.

Chapman, P.M., Dexter, R.N., Anderson, H.B., and Power, E.A., Evaluation of effects associated with an oil platform, using the sediment quality criteria, *Environ. Toxicol. Chem.*, 10, 407, 1991.

Chapman, P.M., Power, E.A., and Burton, G.A., Jr., Integrative assessments in aquatic ecosystems, in *Sediment Toxicity Assessment*, Burton, G.A., Ed., Lewis Publishers, Boca Raton, FL, 1992, 313.

Chen, W. and Morrison, G., Bacterial enzyme activity and metal speciation in urban river sediments (manuscript), Chalmers University of Gothenburg, Sweden, 1991.

Chester, R. and Hughes, M.J., A chemical technique for the separation of ferromanganese minerals, carbonate minerals, and adsorbed trace elements from pelagic sediments, *Chem. Geol.*, 2, 249, 1967.

Christensen, T.H., Cadmium soil sorption at low concentrations. I. Effect of time, cadmium load, pH and calcium, *Water Air Soil Pollut.*, 21, 105, 1984.

Davis, J.A., Complexation of trace metals by adsorbed natural organic matter, *Geochim. Cosmochim. Acta*, 48, 679, 1984.

Davis, J.A. and Kent, D.B., Surface complexation modelling in aqueous geochemistry, in *Mineral–Water Interface Geochemistry*, Reviews in Mineralogy, Vol. 23, Hochella, M.F. and White, A.F., Eds., Mineralogical Society of America, Washington, D.C., 1990, 117.

Diks, D.M. and Allen, H.E., Correlation of copper distribution in a freshwater-sediment system to bioavailability, *Bull. Environ. Contam. Toxicol.*, 30, 37, 1983.

Di Toro, D.M., Mahony, J.D., Hansen, D.J., Scott, K.J., Hicks, M.B., Mayr, S.M., and Redmond, M.S., Toxicity of cadmium in sediments: the role of acid volatile sulfides, *Environ. Toxicol. Chem.*, 9, 1487, 1990.

Di Toro, D.M., Zarba, C.S., Hansen, D.I., Berry, W.J., Swartz, R.C., Cowan, C.E., Pavlou, S.P., Allen, H.E., Thomas, N.A., and Paquin, P.R., Technical basis for establishing water quality criteria for nonionic organic chemicals using equilibrium partitioning, *Environ. Toxicol. Chem.*, 10, 1541, 1991.

Dobbs, J.C., Susetyo, W., Knight, F.E., Castles, M.A., Carreira, L.A., and Azarraga, L.V., Characterization of metal binding sites in fulvic acids by lanthanide ion probe spectroscopy, *Anal. Chem.*, 61, 483, 1989.

Eighmy, T.T., Eusden, J.D., Jr., Krzanowski, J.E., Domingo, D.S., Stämpfli, D., Martin, J.R., and Erickson, P.M., Comprehensive approach toward understanding element speciation and leaching behaviour in municipal solid waste incineration electrostatic precipitator ash, *Environ. Sci. Technol.*, 29, 629, 1995.

Elsokkary, I.H., Amer, M.A., and Shalaby, E.A., Assessment of inorganic lead species and total organo-alkyl lead in some Egyptian agricultural soils, *Environ. Pollut.*, 87, 225, 1995.

Faust, B.C., The octanol/water distribution coefficients of methylmercury species: the role of aqueous-phase chemical speciation, *Environ. Toxicol. Chem.*, 11, 1373, 1992.

Fawaris, B.H. and Johanson, K.J., Fractionation of caesium (^{137}Cs) in coniferous forest soil in central Sweden, *Sci. Total Environ.*, 170, 221, 1995.

Förstner, U., Chemical forms and reactivities of metals in sediments, in *Chemical Methods for Assessing Bioavailable Metals in Sludges and Soils*, Leschber, R., Davis, R.D., and L'Hermite, P., Eds., Elsevier, London, 1985, 1.

Förstner, U., Sediment-associated contaminants — an overview of scientific bases for developing remedial options, *Hydrobiologia*, 149, 221, 1987.

Förstner, U., Non-linear release of metals from aquatic sediments, in *Biogeodynamics of Pollutants in Soils and Sediments*, Salomons, W. and Stigliani, W.M., Eds., Springer-Verlag, Berlin, 1995, 247.

Förstner, U. and Carstens, A., *In-Situ* Versuche zur Veränderung von festen Schwermetallphasen in aeroben und anaeroben Grundwasserleitern, *Vom Wasser*, 71, 113, 1988.

Förstner, U. and Kersten, M., Assessment of metal mobility in dredged material and mine waste by pore water chemistry and solid speciation, in *Chemistry and Biology of Solid Waste,* Salomons, W. and Förstner, U., Eds., Springer-Verlag, Berlin, 1988, 214.

Förstner, U. and Müller, G., *Schwermetalle in Flüssen und Seen,* Springer-Verlag, Berlin, 1974.

Förstner, U. and Schoer, J., Diagenesis of chemical associations of ^{137}Cs and other artificial radionuclides in river sediments, *Environ. Technol. Lett.,* 5, 295, 1984.

Förstner, U. and Wittman, G.T.W., *Metal Pollution in the Aquatic Environment,* Springer-Verlag, Berlin, 1981.

Förstner, U., Ahlf, W., Calmano, W., Schuhmann, C., and Sellhorn, C., Einfluss von NTA auf die Sorption von Schwermetallen an definierten Feststoffphasen (Calcit, Illit, Montmorillonit, Algenzellwände), *Vom Wasser,* 61, 155, 1983.

Förstner, U., Calmano, W., Ahlf, W., and Kersten, M., Ansätze zur Beurteilung der "Sedimentqualität" in Gewässern, *Vom Wasser*, 73, 25, 1989a.

Förstner, U., Kersten, M., and Wienberg, R., Geochemical processes in landfills, in *The Landfill — Reactor and Final Storage,* Baccini, P., Ed., Lecture Notes in Earth Science, Vol. 20, Springer-Verlag, Berlin, 1989b, 39.

Förstner, U., Calmano, W., and Kersten, M., Sediment criteria development — contributions from environmental geochemistry to water quality management, in *Sediments and Environmental Geochemistry,* Heling, D., Rothe, P., Förstner, U., and Stoffers, P., Eds., Springer-Verlag, Berlin, 1990, 311.

Froelich, P.N., Klinkmann, G.P., Beuder, M.L., Leudtke, N.A., Heath, G.R., Cullen, D., Dauphin, P., Hammond, D., Hartman, B., and Maynard, V., Early oxidation of organic matter in pelagic sediments of the eastern equatorial Atlantic: suboxic diagenesis, *Geochim. Cosmochim. Acta,* 43, 1075, 1979.

Goldberg, E.D., Marine geochemistry: chemical scavengers of the sea, *J. Geol.,* 62, 249, 1954.

Goldberg, E.D. and Arrhenius, G.O.S., Chemistry of Pacific pelagic sediments, *Geochim. Cosmochim. Acta,* 13, 153, 1958.

Griffith, S.M. and Schnitzer, M., The isolation and characterization of stable metal–organic complexes from tropical volcanic soils, *Soil Sci.,* 120, 126, 1975.

Gunn, A.M., Hunt, D.T.E., and Winnard, D.A., The effect of heavy metal speciation in sediment on bioavailability to tubificid worms, *Hydrobiologia,* 188/189, 487, 1989.

Guy, R.D. and Chakrabarti, C.L., Distribution of metal ions between soluble and particulate forms, in Abstr. Int. Conf. Heavy Metals in the Environment, Toronto, 1975.

Håkanson, L., A bottom sediment trap for recent sedimentary deposits, *Limnol. Oceanogr.,* 21, 170, 1976.

Hall, G.E.M., Determination of trace elements in sediments, in *Manual of Physico-Chemical Analysis of Aquatic Sediments,* Mudroch, A., Azcue, J.M., and Mudroch, P., Eds., Lewis Publishers, Boca Raton, FL, 1996, 85.

Hawkes, H.E. and Webb, J.S., *Geochemistry in Mineral Exploration,* Harper and Row, New York, 1962, 415.

Hendrickson, L.L. and Corey, R.B., A chelating resin method for characterizing soluble metal complexes, *Soil Sci. Soc. Am. J.,* 47, 467, 1983.

Herms, U. and Bruemmer, G., Einfluß der Bodenreaktion auf Löslichkeit und tolerierbare Gesamtgehalte an Nickel, Kupfer, Zink, Cadmium und Blei in Böden und kompostierten Siedlungsabfällen, *Landwirtsch. Forsch.*, 33, 408, 1980.

Heron, G. and Christensen, T.H., Impact of sediment-bound iron on redox buffering in a landfill leachate polluted aquifer (Vejen, Denmark), *Environ. Sci. Technol.*, 29, 187, 1995.

Heron, G., Crouzet, K., Bourg, A.C.M., and Christensen, T.H., Speciation of Fe(II)/Fe(III) in contaminated aquifer sediments using chemical extraction techniques, *Environ. Sci. Technol.*, 28, 1698, 1994.

Hesslein, R.H., An *in-situ* sampler for close interval pore water studies, *Limnol. Oceanogr.*, 21, 912, 1976.

Hirner, A.V., Trace element speciation in soils and sediments using sequential chemical extraction methods, *Int. J. Environ. Chem.*, 46, 77, 1992.

Hirner, A.V., Testing metal mobility in soils by elution tests, in *Geochemical Approaches for Environmental Engineering of Metals*, Reuther, R., Ed., Springer-Verlag, Berlin, 1996.

Hirner, A.V. and Xu, Z., Trace metal speciation in Julia Creek oil shale, *Chem. Geol.*, 91, 115, 1991.

Honeyman, B.D. and Santschi, P.H., Metals in aquatic systems — predicting their scavenging residence times from laboratory data remains a challenge, *Environ. Sci. Technol.*, 22, 862, 1988.

Huang, C.P., The surface acidity of hydrous solids, in *Adsorption of Inorganics at Solid–Liquid Interfaces*, Anderson, M.A. and Rubin, A.J., Eds., Ann Arbor Science, Ann Arbor, MI, 1981.

Ingersoll, C.G., Brumbaugh, W.G., Dwyer, F.J., and Kemble, N.E., Bioaccumulation of metals by *Hyalella azteca* exposed to contaminated sediments from the Upper Clark River, Montana, *Environ. Toxicol. Chem.*, 13, 2013, 1994.

Irving, H. and Williams, R., Order of stability of metal complexes, *Nature*, 162, 746, 1948.

Jackson, M.L., *Soil Chemical Analysis*, Prentice-Hall, Englewood Cliffs, NJ, 1958, 498.

James, R.O. and Healy, T.W., Adsorption of hydrolyzable metal ions at the oxide–water interface, *J. Colloid Interface Sci.*, 40, 42, 1972.

James, B.R., Petura, J.C., Vitale, R.J., and Mussoline, G.R., Hexavalent chromium extraction from soils: a comparison of five methods, *Environ. Sci. Technol.*, 29, 2377, 1995.

Jenkins, R. and de Vries, J.L., *An Introduction to X-ray Powder Diffractometry*, Eindhoven, Netherlands, 1970.

Jenne, E.A., Trace element sorption by sediments and soils, in *Sites and Processes, Proc. Symp. Molybdenum in the Environment*, Vol. 2, Chappel, W. and Peterson, K.S., Eds., Dekker, New York, 1977, 425.

Jenne, E.A. and Luoma, S.N., Forms of trace elements in soils, sediments and associated waters: an overview of their determination and biological availability, in Proc. 5th Annu. Hanford Life Sciences Program, Ritchland, VA, 1977.

Johnson, C.A., Brandenberger, S., and Baccini, P., Acid neutralizing capacity of municipal waste incinerator bottom ash, *Environ. Sci. Technol.*, 28, 142, 1995.

Jonasson, I.R., Geochemistry of sediment/water interactions of metals, including observations on availability, in *The Fluvial Transport of Sediment-Associated Nutrients and Contaminants*, Shear, H. and Watson, A.E.P., Eds., IJC/PLUARG, Windsor, ON, 1977, 255.

Kemble, N.E., Brumbaugh, W.G., Brunson, E.L., Dwyer, F.J., Ingersoll, C.G., Monda, D.P., and Woodward, D.F., *Environ. Toxicol. Chem.*, 13, 1985, 1994.

Kersten, M. and Förstner, U., Chemical fractionation of heavy metals in anoxic estuarine and coastal sediments, *Water Sci. Technol.*, 18, 121, 1986.

Kersten, M. and Förstner, U., Effect of sample pretreatment on the reliability of solid speciation data of heavy metals — implications for the study of early diagenetic processes, *Mar. Chem.*, 22, 299, 1987.

Keyser, T.R., Natush, D.F.S., Evans, C.A., and Linton, R.W., Characterizing the surface of environmental particles, *Environ. Sci. Technol.*, 12, 768, 1978.

Kheboian, C. and Bauer, C.F., Accuracy of selective extraction procedures for metal speciation in model aquatic sediments, *Anal. Chem.*, 59, 1417, 1987.

Koch, O.G. and Koch-Dedic, G.A., *Handbuch der Spurenanalyse*, Springer-Verlag, Berlin, 1974, 1232.

Kong, I.-C. and Liu, S.-M., Determination of heavy metals distribution in the anoxic sediment slurries by chemical sequential fractionation, *Ecotoxicol. Environ. Saf.*, 32, 34, 1995.

Krantzberg, G., Spatial and temporal variability in metal bioavailability and toxicity of sediment from Hamilton Harbour, Lake Ontario, *Environ. Toxicol. Chem.*, 13, 1685, 1994.

Kwan, K.K. and Dutka, B.J., Comparative assessment of two solid-phase toxicity bioassays: the direct sediment toxicity testing procedure (DSTTP) and the Microtox solid-phase test (SPT), *Bull. Environ. Contam. Toxicol.*, 55, 338, 1995.

Landner, L., *Speciation of Metals in Water, Sediment and Soil Systems*, Lecture Notes in Earth Sciences, Springer-Verlag, Berlin, 1987, 190.

Leonard, E.N., Mattson, V.R., and Ankley, G.T., Horizon-specific oxidation of acid volatile sulfide in relation to the toxicity of cadmium spiked into a freshwater sediment, *Arch. Environ. Contam. Toxicol.*, 28, 78, 1995.

Leschber, R., Davis, R.D., and L'Hermite, P., *Chemical Methods for Assessing Bio-Available Metals in Sludges and Soils*, Elsevier, Amsterdam, 1985, 104.

Losi, M.E., Amrhein, C., and Frankenberger, W.T., Jr., Factors affecting chemical and biological reduction of hexavalent chromium in soil, *Environ. Toxicol. Chem.*, 13, 1727, 1994.

Luoma, S.N., Bioavailability of trace metals to aquatic organisms — a review, *Sci. Total Environ.*, 28, 1, 1983.

Luoma, S.N. and Bryan, G.W., Factors controlling the availability of sediment-bound lead to the estuarine bivalve *Scrobicularia plana*, *J. Mar. Biol. Assoc. U.K.*, 58, 793, 1978.

Luoma, S.N. and Bryan, G.W., Trace metal bioavailability: modelling chemical and biological interactions of sediment-bound Zn, in *Chemical Modelling in Aqueous Systems*, Jenne, E.A., Ed., ACS Symp. Ser. 93, American Chemical Society, Washington, D.C., 1979, 577.

Luoma, S.N. and Bryan, G.W., A statistical study of environmental factors controlling concentrations of heavy metals in the burrowing bivalve *Scrobicularia plana* and the polychaete *Nereis diversicolor*, *Estuarine Coastal Shelf Sci.*, 15, 95, 1982.

Maienthal, E.J. and Becker, D.A., A survey of current literature on sampling, sample handling for environmental material, and long-term storage, *Interface*, 5, 49, 1976.

Malo, B.A., Partial extraction of metals from aquatic sediments, *Environ. Sci. Technol.*, 11, 277, 1977.

Martin, J.M., Nirel, P., and Thomas, A.J., Sequential extraction techniques: promises and problems, *Mar. Chem.*, 22, 313, 1987.

Martin, N., Schuster, I., and Peiffer, St., Two experimental methods to determine the speciation of cadmium in sediment from the River Neckar, *Acta Hydrochim. Hydrobiol.*, 24, 68, 1996.

Maxwell, J.A., *Rock and Mineral Analysis*, Interscience, New York, 1968, 506.

Mayer, L.M., Chemical water sampling in lakes and sediments with dialysis bags, *Limnol. Oceanogr.*, 21, 909, 1976.

Meyer, J.S., Davidson, W., Sundby, W., Oris, J.T., Lauren, D.J., Förstner, U., Hong, J., and Crosby, D.G., The effects of variable redox potentials, pH and light on bioavailability in dynamic water–sediment environments, in *Bioavailability — Physical, Chemical, Biological Interactions*, Hamelink, J., Landrum, P.F., Bergman, H.L., and Benson, W.H., Eds., Lewis Publishers, Boca Raton, FL, 1994, 155.

Microbics Corporation, *Microtox Manual: A Toxicity Testing Handbook*, Vol. 2, Carlsbad, CA, 1992.

Moore, J.N., Ficklin, W.H., and Johns, C., Partitioning of arsenic and metals in reducing sulfidic sediments, *Environ. Sci. Technol.*, 22, 432, 1988.

Morrison, G.M.P., Batley, G.E., and Florence, T.M., Metal speciation and toxicity, *Chem. Br.*, p. 791, August 1989.

Mudroch, A. and Azcue, J.M., *Manual of Aquatic Sediment Sampling*, Lewis Publishers, Boca Raton, FL, 1995, 219.

Mudroch, A. and MacKnight, S.D., *Handbook of Techniques for Aquatic Sediments Sampling*, 2nd ed., Lewis Publishers, Boca Raton, FL, 1994, 236.

Müller, G., *Methoden der Sedimentuntersuchungen*, Schweizerbart, Stuttgart, Germany, 1964.

Müller, G., Kretzer, W., and Hirner, A., Zur Methodik von Schwebstoffuntersuchungen an Flußwässern, *GWF Wasser Abwasser*, 117, 220, 1976.

Nelson, A. and Donkin, P., Processes of bioaccumulation: the importance of chemical speciation, *Mar. Pollut. Bull.*, 16, 164, 1985.

Neubecker, T.A. and Allen, H.E., The measurement of complexation capacity and conditional stability constants for ligands in natural water: a review, *Water Res.*, 17, 1, 1983.

Nordstrom, D.K., Plummer, L.N., Wigley, T.L.M., Wolery, T.J., Ball, J.W., Jenne, E.A., Bassett, R.L., Crerar, D.A., Florence, T.M., Fritz, B., Hoffman, M., Holdren, G.R., Lafon, G.M., Jr., Mattigod, S.V., McDuff, R.E., Morel, F., Reddy, M.M., Sposito, G., and Thraikill, J., Comparison of computerized chemical models for equilibrium calculations in aqueous systems, in *Chemical Modelling in Aqueous Systems: Speciation, Sorption, Solubility and Kinetics*, Jenne, E.A., Ed., Symp. Ser. 93, American Chemical Society, Washington, D.C., 1979.

Norsih, K., *Trace Elements in Soil–Plant–Animal Systems*, Academic Press, New York, 1975, 55.

Novotny, V., Diffuse sources of pollution by toxic metals and impact on receiving waters, in *Heavy Metals*, Salomons, W., Förstner, U., and Mader, P., Eds., Springer-Verlag, Berlin, 1995, 33.

Obermann, P. and Cremer, S., Mobilisierung von Schwermetallen in Porenwässern von belasteten Böden und Deponien: Entwicklung eines aussagekräftigen

Elutionsverfahrens, Materialien zur Ermittlung und Sanierung von Altlasten, Landesamt Wasser und Abfall Nordrhein-Westfallen, Bd. 6, 1992.

Orth, R.G., Powell, R.L., Kutey, G., and Kimerle, R.A., Impact of sediment partitioning methods on environmental safety assessment of surfactants, *Environ. Toxicol. Chem.*, 14, 337, 1995.

Pacyna, J.M. and Münch, J., Anthropogenic mercury emission in Europe, *Water Air Soil Pollut.*, 56, 51, 1991.

Pesch, C.E., Hansen, D.J., Boothman, W.S., Berry, W.J., and Mahony, J.D., The role of acid-volatile sulfide and interstitial water metal concentrations in determining bioavailability of cadmium and nickel from contaminated sediments to the marine polychaete *Neanthes arenaceodentata, Environ. Toxicol. Chem.*, 14, 129, 1995.

Pickering, W.F., Metal ion speciation — soils and sediments (a review), *Ore Geol. Rev.*, 1, 83, 1986.

Plant, J.A. and Raiswell, R., Principles of environmental geochemistry, in *Applied Environmental Geochemistry*, Thornton, I., Ed., Academic Press, London, 1983, 1.

Rapin, F., Nembrini, G.P., Förstner, U., and Garcia, J.I., Heavy metals in marine sediment phases determined by sequential chemical extraction and their interaction with interstitial water, *Environ. Technol. Lett.*, 4 387, 1983.

Rashid, M.A., Role of humic acids of marine origin and their different molecular weight fractions in complexing di- and trivalent metals, *Soil Sci.*, 111, 298, 1971.

Rashid, M.A., Adsorption of metals on sedimentary and peat humus acids, *Chem. Geol.*, 13, 115, 1974.

Rendell, P.S., Batley, G.E., and Cameron, A.J., Adsorption as a control of metals concentrations in sediment extracts, *Environ. Sci. Technol.*, 14, 314, 1980.

Reuther, R., A Literature Review on the Occurrence and Speciation of Arsenic in the Aquatic Environment, MFG technical report, 1986.

Reuther, R., Arsenic introduced into a littoral freshwater model ecosystem, *Sci. Total Environ.*, 115, 219, 1992.

Reuther, R., *Geochemical Approaches for Environmental Engineering of Metals*, Springer-Verlag, Berlin, 1996.

Reuther, R. and Grahn, O., Chemical forms of aluminium and their relevance for the turnover of heavy metals in acid, limed and metal-polluted lake sediments, in Proc. Int. Conf. Heavy Metals in the Environment, Athens, 1985, 443.

Reuther, R., Wright, R.F., and Förstner, U., Distribution and chemical forms of heavy metals from two Norwegian lakes, affected by acid precipitation, in Proc. Int. Conf. Heavy Metals in the Environment, Amsterdam, 1981, 318.

Robins, J.M., Lyle, M., and Heath, G.R., A Sequential Extraction Procedure for Partitioning Elements Among Coexisting Phases in Marine Sediments, College of Oceanography Report 84-3, Oregon State University, Corvallis, 1984.

Rosa, F., Bloesch, J., and Rathke, D.E., Sampling the settling and suspended particulate matter (SPM), in *Handbook of Techniques for Aquatic Sediment Sampling*, Mudroch, A. and MacKnight, S.D., Eds., Lewis Publishers, Boca Raton, FL, 1994, 97.

Rose, A.W., The mode of occurrence of trace elements in soils and stream sediments applied to geochemical exploration, in *Geochemical Exploration*, Elliot, I.L. and Fletcher, W.K., Eds., Elsevier, Amsterdam, 1975.

Rose, A.W., Hawkes, H.E., and Webb, J.S., *Geochemistry in Mineral Exploration*, Academic Press, London, 1979.

Salomons, W., Adsorption processes and hydrodynamic conditions in estuaries, *Environ. Technol. Lett.*, 1, 356, 1980.

Salomons, W., Long-term strategies for handling contaminated sites and large-scale areas, in *Biogeodynamics of Pollutants in Soils and Sediments*, Salomons, W. and Stigliani, W.M., Eds., Springer-Verlag, Berlin, 1995, 1.

Salomons, W. and Förstner, U., Trace metal analysis of polluted sediment. II. Evaluation of environmental impact, *Environ. Technol. Lett.*, 1, 506, 1980.

Salomons, W. and Förstner, U., *Metals in the Hydrocycle*, Springer-Verlag, Berlin, 1984, 349.

Salomons, W. and Stigliani, W.M., *Biogeodynamics of Pollutants in Soils and Sediments*, Springer-Verlag, Berlin, 1995, 352.

Salomons, W., Förstner, U., and Mader, P., *Heavy Metals: Problems and Solutions*, Springer-Verlag, Berlin, 1995, 412.

Saradin, P.M., Lapaquellerie, Y., Astruc, A., Latouche, C., and Astruc, M., Long-term behaviour and degradation kinetics of tributyltin in a marine sediment, *Sci. Total Environ.*, 170, 59, 1995.

Sauerbeck, D.R. and Styperek, P., Evaluation of chemical methods for assessing the Cd and Zn availability from different soils and sources, in *Chemical Methods for Assessing Bioavailable Metals in Sludges and Soils*, Leschber, R. et al., Eds., Elsevier Applied Science, New York, 1985, 49.

Schoer, J. and Förstner, U., Abschätzung der Langzeitbelastung von Grundwasser durch die Ablagerung metallhaltiger Feststoffe, *Vom Wasser*, 69, 23, 1987.

Shea, D., Developing national sediment quality criteria — equilibrium partitioning of contaminants as a means of evaluating sediment quality criteria, *Environ. Sci. Technol.*, 22, 1256, 1988.

Sobek, A.A., Schuller, W.A., Freeman, J.R., and Smith, R.M., Field and Laboratory Methods Applicable to Overburden and Mine Spoils, Report EPA-600/2-78-054, U.S. Environmental Protection Agency, Washington, D.C., 1978.

Stumm, W. and Morgan, J.J., *Aquatic Chemistry*, Wiley, New York, 1981, 780.

Suedel, B.C. and Rodgers, J.H., Jr., Development of formulated reference sediments for freshwater and estuarine sediment testing, *Environ. Toxicol. Chem.*, 13, 1163, 1994.

Surija, B. and Branica, M., Distribution of Cd, Pb, Cu and Zn in carbonate sediments from the Krka River estuary obtained by sequential extraction, *Sci. Total Environ.*, 170, 101, 1995.

Swift, R.S., *Soil Organic Matter Studies*, IAEA, Vienna, 1977, 275.

Tay, K.L., Doe, K.G., Wade, S.J., Vaughan, D.A., Berrigan, R.E., and Moore, M.J., Sediment bioassessment in Halifax Harbour, *Environ. Toxicol. Chem.*, 11, 1567, 1992.

Tessier, A., Campbell, P.G.C., and Bisson, M., Sequential extraction procedure for the speciation of particulate trace metals, *Anal. Chem.*, 51, 844, 1979.

Tessier, A., Campbell, B.G.C, Auclair, J.C., and Bisson, M., Relationship between the partitioning of trace metals in sediments and their accumulation in the tissues of the freshwater mollusc *Elliptio complanata* in a mining area, *Can. J. Fish. Aquat. Sci.*, 41, 1463, 1984.

Trefry, J.H. and Metz, S., Selective leaching of trace metals from sediments as a function of pH, *Anal. Chem.*, 56, 745, 1984.

Turner, D.R. and Whitfield, M., Chemical definition of the biological available fraction of trace metals in natural waters, *Thalassia Jugosl.*, 16, 227, 1980.

Ure, A.M., Quevauviller, P.M., Muntau, M., and Griepink, B., Speciation of heavy metals in soils and sediments: an account of the improvement and harmonization of extraction techniques undertaken under the auspices of the BCR of the Commission of the European Communities, *Int. J. Environ. Anal. Chem.*, 51, 135, 1993.

U.S. Environmental Protection Agency, Hazardous waste management. I. Toxicity characteristic leaching procedure, *Fed. Reg.*, 51, 02-1766, 1986.

Westall, J., Adsorption mechanisms in aquatic surface chemistry, in *Aquatic Surface Chemistry*, Stumm, W., Ed., Wiley-Interscience, New York, 1987, 3.

Westall, J., Zachary, J.L., and Morel, F.M.M., *MINEQL: A Computer Program for the Calculation of Chemical Equilibrium Composition of Aqueous Systems*, MIT, Cambridge, MA, 1976.

Whitfield, M. and Turner, D.R., Chemical periodicity and the speciation and cycling of the elements, in *Trace Metals in Sea Water*, Wong, C.S., Boyle, E., Bruland, K.W., Burton, J.D., and Goldberg, E.D., Eds., Plenum Press, New York, 1983, 719.

Zhuang, Y., Allen, H.E., and Fu, G., Effect of aeration on cadmium binding, *Environ. Toxicol. Chem.*, 13, 717, 1994.

Züllig, W., Sedimente als Ausdruck des Zustandes eines Gewässers, *Schweiz. Z. Hydrol.*, 18, 7, 1956.

chapter two

Sediment bacterial populations and methodologies to study the populations important to xenobiotic degradation

Gordon Southam and Colette MacKenzie

2.1 *Introduction*

Microbial activity in sediments is typically enhanced compared to that of the planktonic bacterial populations inhabiting the hydrosphere above the sediments. Therefore, the populations of bacteria in sediments have the most profound effects on redox reactions, nutrient cycling, and the bioremediation/transformation of toxic materials. By asking questions pertaining to the populations and their source(s) of carbon and energy, their terminal electron acceptor(s), and by-products of their metabolism, we can begin to understand the effects these organisms have on their environments.

One of the most interesting features of sediments is their ability to function as a "sink" for contaminants which enter associated aquatic habitats (Ferris et al., 1987). In the study of bacteria in sediments, the foremost consideration is their microenvironment. Due to the different sizes, bacteria and sediment investigators often have very different perceptions of the environments or scale in which bacteria function. Therefore, the reactions catalyzed by bacteria in the natural environment often do not coincide with relatively macroscopic chemical measurements/gradients such as pH and dissolved oxygen (Enzien et al., 1994). Tremendous advances in the study of the ecology of microorganisms and their microenvironments have been achieved through the use of microprobes, which enable measurement of pH, O_2, and H_2S gradients over distances smaller than 1 mm (Revsbech and Jorgensen, 1986; Emerson and Revsbech, 1994a,b; Jensen et al., 1994). While

1-56670-343-3/99/$0.00+$.50

this technique also possesses tremendous analytical precision, microprobe analysis cannot be applied to routine sediment work due to its high cost and extremely delicate nature.

As new technologies emerge, researchers tend to abandon traditional methods, assuming that "newer is better." Since the introduction of the polymerase chain reaction (PCR) in 1987, there has been an explosion in its use. We suggest, however, that PCR should have limited application in the study of populations in the natural environment. Its prohibitive cost and extreme sensitivity do not lend themselves to the examination of dominant sediment microbial populations. While no one can dispute the incalculable importance of the ability of PCR to detect a single environmental pathogen, its detection of a single sulfate-reducing bacterium in a gram of sediment is not itself a significant finding.

We have chosen to focus on traditional analyses such as plate counting and most probable number analysis using selective and differential growth media. Focus will be primarily on the phenotypic expressions of bacteria which can be measured via the cultivation of specific groups of bacteria. The classical methods are exceedingly effective for the examination of natural microbial populations in sediments. Systematic approaches will be offered which can be undertaken by the novice microbiologist with minimal investment in equipment and training. The physico-chemical analysis of sediments can be greatly enhanced by the incorporation of these basic techniques. The small investment will yield a multifaceted perspective which more truly reflects the sediment system.

2.2 Literature review

Bacteria are a ubiquitous group of prokaryotic organisms which exhibit such metabolic diversity that they first must be categorized before their study can even be attempted. *Bergey's Manual of Systematic Bacteriology* (Holt, 1984) is a four-volume series that represents the most complete guide to bacterial systematics. *Bergey's Manual* groups bacteria according to morphology, physiology, and ecology. The focus here will, in part, be on the second and primarily the third criteria, thereby grouping organisms according to impact on their environment or physiological criteria related to "what they do."

Bacterial processes are established through competition for available nutrients and through the efficiency of their respective energy generation (nutrient utilization) mechanisms. In most aquatic environments, the water/sediment interface is an aerobic environment in which oxygen serves as the terminal electron acceptor. Oxygenic photosynthesis may also be active within this region and may manifest itself as algal or cyanobacterial mats. Below this region of oxygenic photosynthesis and aerobic mineralization, anaerobic processes are stratified. The stratification is controlled by the energy-generating efficiency of each successive terminal electron acceptor (Lovley

Water

$Fe^{3+} + OH^- \rightarrow Fe(OH)_{3(s)}$

- - - - - - Water/Sediment Interface - - - - -

$(CH_2O)_n + O_2 \rightarrow CO_2$

$HCO_3^- + H_2O \rightarrow O_2 + *(CH_2O)_n$

$HCO_3^- + Fe^{2+} + O_2 \rightarrow *(CH_2O)_n + Fe^{3+}$

$HCO_3^- + NH_4^+ + O_2 \rightarrow *(CH_2O)_n + NO_3^-$

Sediment

- - - - - - Aerobic/anaerobic interface - - - -

$(CH_2O)_n + NO_3^- \rightarrow (CH_2O)_{n-x} + CO_2 + N_{2(g)}$

$(CH_2O)_n + Fe(OH)_{3(s)} \rightarrow (CH_2O)_{n-x} + CO_2 + Fe^{2+}$

$(CH_2O)_n + MnO_{2(S)} \rightarrow (CH_2O)_{n-x} + CO_2 + Mn^{2+}$

$(CH_2O)_n + SO_4^{2-} \rightarrow (CH_2O)_{n-x} + CO_2 + HS^-$

$HCO_3^- + H_2O \rightarrow CH_4 + *(CH_2O)_n$

Figure 2-1 Schematic representation of bacterial populations associated with nonstratified aquatic sediments. The bacterial groups are represented by their basic energetic processes. This list of physiological groups is not complete but highlights the stratification which exists in all sediment environments based on the energetics of the microbial processes. $(CH_2O)_n$ denotes organic carbon (source of carbon for biomass and for energy generation); $(CH_2O)_{n-x}$ denotes incomplete oxidation of organic carbon; $*(CH_2O)_n$ denotes the formation of organic carbon for biomass (autotrophs).

and Klug, 1983) (refer to Figure 2-1). In stratified lakes, the aerobic/anaerobic interface occurs within the thermocline (during the summer months), and oxygenic photosynthesis will not occur on the "surface" of the sediment. In these systems, fermentative processes will predominate throughout the sediment environment. In shallow aquatic environments that con-

Table 2-1 Classification Schemes of Bacteria

Physico-chemical constraints	Microbial physiology
Substratum	Source of carbon
Temperature	Source of energy
Oxygen	Terminal electron acceptors
pH	
Dissolved organic compounds	
Dissolved inorganic compounds	
Water availability	

tain relatively high levels of primary productivity, such as bogs and swamps, the metabolism of organic carbon will also occur, primarily by bacteria through anaerobic processes. Both aerobic and anaerobic bacteria are important components of the food chain because they utilize dilute concentrations of inorganic and organic nutrients that are unavailable to higher eukaryotic organisms (Fenchel and Jorgensen, 1977).

Classification schemes of bacteria based on different environmental conditions and microbial physiology are presented in Table 2-1.

2.2.1 Classification schemes based on physico-chemical constraints

Bacteria can be classified according to the effect that various physical and chemical environments have on their growth and survival. Understanding these environmental influences allows us to understand the distribution of microorganisms in the natural environment. It also enables us to differentially select target organisms by attempting to mimic their preferred growth environments in the laboratory.

2.2.1.1 Substratum

The growth of bacteria in the natural environment typically occurs as biofilms (Konhauser et al., 1994). While biofilms often refer to thick, porous layers of bacteria (>100 μm thick) encased in exopolysaccharide material (capsule or slime layers) (Korber et al., 1995), biofilms will be defined here as particle-associated (as opposed to free-living) prokaryotes (Fukui and Takii, 1990). In the sediment environment, a biofilm includes the colonization of inorganic particles (Fukui and Takii, 1990) as small as 10 μm (Southam et al., 1995). Bacterial associations with sediment particles may be initiated by ionic interactions (Mills et al., 1994) or hydrophobicity (Chakrabarti and Banerjee, 1991).

2.2.1.2 Temperature

Temperature is a measurable physical condition that can differentiate bacteria into broad groups. Temperature is also the most reliable parameter in microbial ecology because it is not affected by biogeochemical cycling.

Psychrophiles grow optimally at or below 15°C. Psychrotolerant bacteria grow optimally between 20 and 40°C but will grow at temperatures as low as 0°C. Mesophiles grow optimally between 15 and 45°C. The optimum growth temperature is above 45°C for thermophiles and above 80°C for hyperthermophiles. With the exception of relatively rare hydrothermal systems, sediment environments will often contain psychrophilic or mesophilic microorganisms.

2.2.1.3 Oxygen

The utilization of oxygen as a terminal electron acceptor enables organisms to fully oxidize their respective source(s) of energy and thereby maximize their energy gain from the reaction. Any organism which is able to grow in the presence of oxygen, whether or not it utilizes oxygen as a terminal electron acceptor, must have the capacity to detoxify the chemically reactive oxygen compounds (hydrogen peroxide, superoxide radical, and the hydroxy radical) produced through the stepwise reduction of oxygen to water (Atlas and Bartha, 1993). Detoxification of these oxygen radicals requires a series of enzymes, including superoxide dismutase, catalase, and peroxidase. The ability to tolerate oxygen relates to the presence or absence of each of these enzymes. Measurements of dissolved oxygen and redox potential are important in determining which microbial processes are functioning within a sediment environment.

2.2.1.4 pH

Sediments occur primarily as pH neutral environments that contain hydrogen ion concentrations of 100 μM, plus or minus one order of magnitude (pH 6 to 8). Bacteria (with the exception of acidophiles and alkalophiles) are typically most active in this pH range. However, sediments impacted by acid mine drainage will exist at a lower pH (Gyure et al., 1987), while sediments associated with carbonate-containing host rocks will exist at a higher pH (Southam, 1995). The indigenous bacteria will reflect these extremes in environmental pH (Goodwin and Zeikus, 1987) and will be either acidophilic (acidotolerant) or alkalophilic (alkalotolerant).

2.2.1.5 Dissolved organic compounds

While organic matter is typically mineralized through aerobic and anaerobic processes in the natural environment, some organic compounds such as carboxylic acids, alcohols, and phenolics can inhibit bacterial growth (Atlas and Bartha, 1993). These compounds can accumulate in anaerobic environments where the incomplete mineralization of the organic materials can lead to toxic accumulations.

2.2.1.6 Dissolved inorganic compounds

These compounds include dissolved gases (carbon monoxide, carbon dioxide, hydrogen, nitrogen, and dihydrogen sulfide), soluble cations (sodium, calcium, magnesium, ammonium, ferrous and ferric iron), base metals (chromium, nickel, copper, cobalt, zinc, lead, mercury), and soluble anions (chloride, nitrite, nitrate, hydrogen sulfide, sulfite, sulfate, phosphorus, selenate, and arsenate).

A wide range of inorganic nutrients can be utilized by bacteria. The ionic forms of nitrogen (NH_4^+, NO_2^-, and NO_3^-), sulfur (HS^- and SO_4^{2-}), and phosphorus (PO_4^{3-}) can be utilized in assimilation reactions by bacteria. Among these compounds, nitrite, nitrate, and sulfate can also be used in dissimilatory processes as electron acceptors. The ionic strength of the aqueous system is also an important determination of bacterial sorption to particles in sediments (Mills et al., 1994). In particular, iron oxide coatings are known to promote bacteria–silicate interactions within oxic environments (Urrutia and Beveridge, 1993).

There is an equally large number of inorganic compounds that are toxic to bacteria. In particular, the trace amounts of heavy metals which serve as enzyme co-factors become toxic at relatively low concentrations (Gadd, 1993).

2.2.1.7 *Water availability*

All life on earth is based on a requirement for water. Bacteria strategically obtain water through osmosis, resulting in the formation of cell turgor pressure. The bacterial cell envelope is designed to withstand turgor pressure which measures between 2 and 3 atm in Gram-negative eubacteria and up to 15 atm in Gram-positive eubacteria (Beveridge, 1981). With the exception of hypersaline environments (e.g., Great Salt Lake, Utah), most aquatic and sediment environments possess an abundance of available water, whereas terrestrial environments are frequently water stressed.

2.2.2 *Classification schemes based on microbial physiology*

Basic physiological categories can also be used to group bacteria. In order to emphasize the ability of bacteria to alter their environment, we will concentrate on dissimilatory rather than assimilatory processes. In this section, basic physiological groups will be defined according to carbon and energy requirements and terminal electron acceptors.

2.2.2.1 *Source of carbon*

Carbon is the most important criterion by which bacterial populations are characterized. Autotrophs are able to obtain their cellular carbon for biomass from inorganic sources (dissolved CO_2/HCO_3^-). For their source of carbon and energy, heterotrophs require organic carbon compounds ranging from simple organics such as amino acid asparagine (Goldman and Wilson, 1977) or glucose (most heterotrophic species) to macromolecular materials such as cellulose (Ljungdahl and Eriksson, 1985). Autotrophic organisms direct most of their energy toward the fixation (reduction) of CO_2 into organic carbon and therefore tend to grow at slower rates than do heterotrophic organisms.

The ability of bacteria to metabolize organic compounds in the natural environment is dependent on the nature of the chemical, the composition and mineralogy of the soil (sediment), and the mechanism by which the chemical enters the environment (Knaebel et al., 1994). For example, Scow and Hutson (1992) found significant decreases in degradation of glutamate and phenol in the presence of kaolinite due to the effects of sorption and reduced diffusion from the clay surface.

2.2.2.2 Source of energy

The requirement of energy for biosynthetic reactions may be described according to electron source, electron donor, or source of reducing power. The generation of energy by bacteria can be classified into two main processes: substrate-level phosphorylation and electron transport (Caldwell, 1995).

Phototrophs utilize sunlight to release ATP from otherwise chemically stable reduced compounds. Oxygenic phototrophs use light in combination with H_2O as their source of energy. Anoxygenic phototrophs such as anaerobic green and purple sulfur bacteria derive their energy for growth from light and H_2S, producing elemental sulfur or sulfate as end products of their metabolism.

Chemolithotrophs (hydrogen oxidizers, ammonia oxidizers, sulfur oxidizers, and methylotrophs) are aerobic bacteria which utilize inorganic chemicals (H_2, NH_4^+, H_2S, and CH_4) as their energy sources (electron donors). Lithotrophs grow more slowly than heterotrophs due to the relatively low levels of chemical energy present in inorganic compounds.

Chemoorganotrophs, which utilize organic carbon as their source of reducing power (and carbon), include both aerobic and anaerobic bacteria. Aerobic or facultative bacteria (growing as aerobes) maximize their energy through the complete oxidation of organic carbon to CO_2. Anaerobic metabolism of organic compounds is not as energetically efficient, which results in less overall growth (Caldwell, 1995).

2.2.2.3 Terminal electron acceptors

The terminal electron acceptor used by a bacterium will determine the role that the bacterium will play in biogeochemical cycling. Aerobic heterotrophs (some bacteria, most protozoa, and all eukaryotic organisms) utilize O_2 as their terminal electron acceptor and produce CO_2 as a by-product of their metabolism via the complete oxidation of organic carbon. While anaerobic heterotrophs also use organic carbon for biomass, they employ a range of terminal electron acceptors (see Figure 2-1). Nitrate reducers, which produce N_2 gas, are responsible for denitrification. Dissimilatory iron- (manganese-) reducing bacteria reduce and solubilize ferric iron (manganese) precipitates which form in aerobic surface waters and sediment down to the benthos. This ferrous iron may then be precipitated by HS^-, a by-product of dissimilatory sulfate reduction. In the depths of sediment, the anaerobic autotrophs (methanogens) which predominate produce methane from carbon dioxide.

Further differentiation of these groups, along with descriptions of culture media for the bacterial groups involved in the degradation of xenobiotic contaminants in sediments, is provided in Section 2.5.

2.3 Major groups of sediment bacteria

Major groups of sediment bacteria are summarized in Table 2-2 and described in detail below.

2.3.1 Phototrophic eubacteria

Oxygenic phototrophs (cyanobacteria) can be found employing planktonic and sessile (in the upper sediment fraction) growth strategies within the limnetic zone of rivers and lakes. Anoxygenic phototrophs (green and purple sulfur bacteria) can be found within the anaerobic regions of sediments, even where the penetration of light seems unlikely to occur (Southam, 1995). Both groups possess the ability to use light energy to drive ATP synthesis (Stanier et al., 1981) and subsequently produce the NADPH required for the assimilation of CO_2 into organic compounds.

Table 2-2 Major Groups of Sediment Bacteria

Phototrophic eubacteria	Chemolithotrophic eubacteria	Chemoorganotrophic bacteria	Archaeobacteria
Cyanobacteria	Methylotrophs/ methanotrophs	Aerobic heterotrophic bacteria	Methanogenic bacteria
Green and purple sulfur bacteria	Nitrifying bacteria	Anaerobic heterotrophic bacteria:	Extremely halophilic bacteria
	Iron- and manganese-oxidizing bacteria	• Proton-reducing acetogens	Hyperthermophilic bacteria
	Sulfur-oxidizing bacteria	• Dissimilatory nitrate-reducing bacteria	
	Hydrogen-utilizing bacteria	• Dissimilatory metal-reducing bacteria	
		• Dissimilatory sulfur-reducing bacteria	

2.3.1.1 Cyanobacteria

The enumeration of cyanobacteria by in vitro culture is an extremely tedious exercise as cyanobacteria are both slow growing and difficult to grow. Two possible methods are recommended for the characterization of cyanobacterial populations from sediment environments. The first is a direct microscopic count. Cyanobacteria can be easily enumerated using epifluorescence microscopy because of their inherent fluorescent light-harvesting pigments (Thompson, 1995). Under epifluorescence light microscopy (UV-light) with a red-blue or a green filter, cyanobacteria will appear red or orange, respectively, and can therefore be easily distinguished from background bacteria or sediment particles. The relatively small cell size of cyanobacteria distinguishes them from eukaryotic algae.

The second method which measures $^{14}HCO_3^-$ incorporation (Hubel, 1966) is limited because it measures total CO_2 fixation by cyanobacteria, eukaryotic algae, green and purple sulfur bacteria, and nonphotosynthetic autotrophic bacteria. Photosynthetic and chemolithotrophic incorporation of CO_2 can be differentiated by using light (clear) and dark (opaque or covered) bottles. While this method does not provide a bacterial cell count, it will enable the determination of the level of primary productivity (short-duration experiment) or net primary productivity (using a 24-hr photoperiod experiment) in a sediment environment.

2.3.1.2 Green and purple sulfur bacteria

Anoxygenic photosynthesis is most active in the hypolimnion of stratified lakes. However, limited activity can occur near the sediment–water interface where light and hydrogen sulfide gradients overlap (Southam, 1995). In these environments, bacteria catalyze the anaerobic oxidation of sulfide to elemental sulfur or sulfate using bacteriochlorophyll *a*, which has an absorption spectrum between 720 and 790 nm for green sulfur bacteria and 850 and 880 nm for purple sulfur bacteria.

2.3.2 Chemolithotrophic eubacteria

Chemolithotrophic ("rock-eating") eubacteria are able to oxidize inorganic compounds to generate ATP. Culture conditions for methylotrophs (methanotrophs) will be described (Section 2.5) because of their importance in the degradation of xenobiotic compounds and in global carbon cycling (i.e., negative regulation of the greenhouse effect). The nitrifying, iron- (manganese-) oxidizing, and sulfur-oxidizing bacteria will be described only in terms of their ecological significance.

2.3.2.1 Methylotrophs/methanotrophs

These bacteria are able to oxidize methane and other C_1 compounds, including methanol and methylamines, as their sole energy source and major carbon source. These obligate aerobes are also important in sediment remediation processes via mineralization of halomethanes, chlorinated ethanes (Fogel et al., 1986), and trichloroethylene (Little et al., 1988).

Increased methane fluxes have been measured from wetlands after exposure to increased CO_2 levels (Dacey et al., 1994), and it has been hypothesized that increases in atmospheric CO_2 may lead to significant increases in methane emissions. Therefore, methylotrophs may play an important role in reducing the flux of methane from sediment environments. This group of bacteria typically resides at the interface between aerobic and anaerobic environments. A methylotrophic culture medium is described in Section 2.5.

2.3.2.2 *Nitrifying bacteria*

These bacteria are chemolithotrophs that utilize the energy derived from nitrification (ammonia oxidation) to assimilate CO_2. Ammonia oxidation is a two-step process ($NH_3 \rightarrow NO_2^-$, $NO_2^- \rightarrow NO_3^-$) performed by different bacteria (Focht and Verstraete, 1977).

Nitrification can have a profound effect on nitrogen cycling in agriculturally impacted environments where nitrogen is not a limiting nutrient (Atlas and Bartha, 1993). Ammonia oxidation increases nitrogen mobilization by reducing the amount of NH_4^+ that is bound to clays and increasing the levels of NO_2^- and NO_3^- which are mobile and freely soluble in water. This results in a net loss of available nitrogen to groundwater. The second step in nitrification is inhibited by acidic pH, so that wherever acid conditions coincide with high amounts of agricultural synthetic fertilizer, there may be problems with nitrite toxicity. In these regions, nitrite can be transported from groundwater into drinking water sources, where nitrite can react chemically with amino compounds to form nitrosamines (carcinogens) and can combine with blood hemoglobin to reduce its oxygen transfer capacity.

Nitrifying bacteria can be grown using the enrichment and plating methods described by Soriano and Walker (1968). Because of their limited involvement in sediment remediation, we will not further elaborate on this group of bacteria.

2.3.2.3 *Iron- and manganese-oxidizing bacteria*

The oxidation of iron ($Fe^{2+} + \frac{1}{4}O_2 + H^+ \rightarrow Fe^{3+} + \frac{1}{2}H_2O$) (Corstjens et al., 1992; Emerson and Revsbech, 1994a,b) and manganese ($Mn^{2+} + \frac{1}{2}O_2 + 2H^+ \rightarrow Mn^{4+} + H_2O$) (Adams and Ghiorse, 1986; Emerson and Ghiorse, 1992) typically occurs within the aerobic/anaerobic interface where the reduced metals encounter an oxidizing environment. This region selects/enriches for these filamentous bacteria because chemical oxidation restricts their growth in aerobic regions. Iron- (manganese-) oxidizing bacteria typically contain sheaths which surround chains of cells. Sheaths of *Sphaerotilis* (Rogers and Anderson, 1976a,b) and *Leptothrix* (Crerar et al., 1979; Adams and Ghiorse, 1987; Boogerd and de Vrind, 1987; Corstjens et al., 1992) are important because they are responsible for an oxidative enzyme-mediated precipitation of these metals. The importance of these bacteria in sediment biogeochemistry relates to their role in the formation of oxidized forms of iron and manganese (minerals). At some point in the future, these minerals will likely serve as electron acceptors for dissimilatory iron- and manganese-reducing bacteria.

Acidophilic iron-oxidizing bacteria (*Thiobacillus* spp.) are most conspicuous in acid mine drainage (AMD) environments, where they impact the surrounding aquatic watershed. This impact is due to by-products of *Thiobacillus* spp. activity which create high concentrations of dissolved sulfate (acidic pH) and toxic heavy metals. Growth of *Thiobacillus* spp. within lake sediments is considered to be restricted due to an oxygen limitation and has resulted in the disposal of sulfidic mine tailings in some lake environments. The optimum recovery of acidophilic iron-oxidizing bacteria is achieved using the ferrous iron-based growth medium described by Silverman and Lundgren (1959) and the most probable number method (Southam and Beveridge, 1992).

2.3.2.4 *Sulfur-oxidizing bacteria*

Sulfur-oxidizing bacteria can be divided into three groups: chemosynthetic colorless (a term which distinguishes them from chlorophyll-containing sulfur bacteria) sulfur-oxidizing bacteria, chemoautotrophic colorless sulfur-oxidizing bacteria (acidophiles common in sulfidic mine tailings, i.e., AMD environments), and photosynthetic sulfur-oxidizing bacteria (see Section 2.3.1.2). Sulfur oxidation results in the formation of sulfuric acid. In sediments, this can lead to the solubilization and mobilization of phosphorus and other mineral nutrients which generally benefit both microorganisms and aquatic plants.

The chemosynthetic sulfur-oxidizing bacteria are microaerophilic organisms. Because sulfide is rapidly oxidized (spontaneously) at neutral pH in the presence of oxygen or Fe^{3+}, they position themselves at the interface of an anaerobic environment (i.e., sediment which is in contact with oxygenated water). In this environment, they derive energy from the partial oxidation of

$$HS^- + \tfrac{1}{2}O_2 \rightarrow S^\circ + H_2O$$

whenever there is an abundant supply of HS^-. The S° which is deposited inside the bacterial cells can be further oxidized (i.e., converted to SO_4^{2-}) when cells are depleted of HS^-.

2.3.2.5 *Hydrogen-utilizing bacteria*

Certain aerobic bacteria are able to utilize hydrogen as well as organic carbon as their energy source. Most anaerobic bacteria also utilize hydrogen, including dissimilatory nitrate-reducing bacteria, dissimilatory ferric iron-(manganese-) reducing bacteria, dissimilatory sulfate-reducing bacteria, methanogens, and homoacetogens ($H_2 + CO_2 \rightarrow$ acetate). Homoacetogenics will not be discussed further because they account for less than 5% of hydrogen transfer (Zeikus et al., 1985).

2.3.3 *Chemoorganotrophic eubacteria*

2.3.3.1 *Aerobic heterotrophic bacteria*

Heterotrophs utilize organic carbon as their source of cell biomass and energy and produce carbon dioxide as a by-product of their metabolism ($C_6H_{12}O_6 + 6O_2 \rightarrow 6CO_2 + 6H_2O$). Phenotypic characterization of heterotrophs is often based on their ability to grow on various organic compounds. Industrial typing scheme sources (e.g., Biolog®) linked to computer databases may utilize as many as 95 different carbon sources for the classification of heterotrophs. Although it is impractical to utilize a single growth medium which incorporates 95 different sources of carbon, general growth media exist which can support the recovery of heterotrophs from environmental samples. For example, Amy et al. (1992) achieved excellent success in recovering diverse populations of heterotrophs from subsurface environments on Difco's R2-A medium (see Section 2.4). The success of the R2-A growth medium likely relates to its wide range of utilizable carbon sources (Figure 2-2).

2.3.3.2 *Anaerobic heterotrophic bacteria*

2.3.3.2.1 *Proton-reducing acetogens.* Under oxygen-limited conditions, facultative organisms and obligately anaerobic bacteria will grow using fermentative metabolism (e.g., proton reduction by acetogens) and an electron acceptor other than to oxygen, such as nitrate/nitrite, ferric iron, sulfate, or HCO_3 (see following sections). Proton-reducing acetogens are found in freshwater sediments that are low in sulfate and H_2 (Atlas and Bartha, 1993). Saturated and unsaturated fatty acids serve as their carbon source. They degrade odd-numbered fatty acids to acetate, propionate, CO_2, and H_2 and carry out β-oxidation of even-numbered-chain fatty acids to acetate and H_2. These bacteria grow on saturated fatty acids only, with low concentrations of H_2 brought about by the presence of either a methanogen or another organism which uses H_2 as an energy source.

H_2 is a stringent feedback inhibitor of hydrogenase and prevents the essential conversion of NADH to NAD. Methanogen growth reduces the H_2 concentration, thereby relieving the inhibitory effects of the hydrogenase. This syntrophic utilization of hydrogen by another organism is called interspecies hydrogen transfer.

2.3.3.2.2 *Dissimilatory nitrate-reducing bacteria.* Nitrate reduction occurs in the absence of oxygen, with NO_3^- serving as the terminal electron acceptor (Figure 2-3b). Nitrate reduction can occur either in the hypolimnion below the thermocline of stratified lakes or in sediments of nonstratified

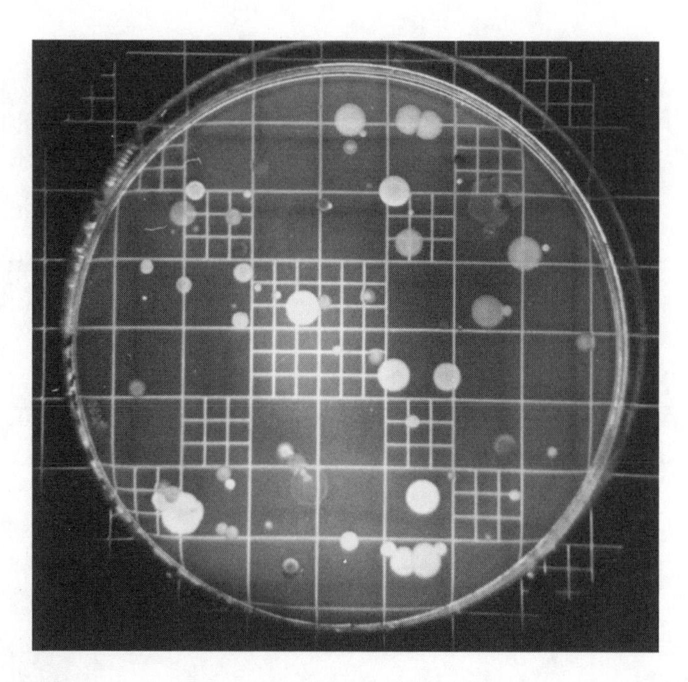

Figure 2-2 Photograph of a petri dish containing R2-A medium which has been inoculated with an appropriate dilution of a coal slurry prepared from Black Mesa coal, Arizona. Note the morphological diversity of the colonies present on the agar surface. Each colony represents a single colony-forming unit even though it is probably derived by more than one bacterium. Statistically significant plates should have between 30 and 300 colonies (inclusive) per plate to be counted.

lakes. Because it occurs at relatively high redox potential, nitrate reduction occurs in the upper zone of the anaerobic layer, above the zones of metal and sulfate reduction (see Figure 2-1).

Denitrification occurs by two pathways. In the first, NO_3^- is reduced to NO_2^-, which is excreted and further converted from NO_2^- to NH_4^+ by other organisms. The second pathway typically results in the complete reduction of NO_3^- ($NO_3^- \rightarrow NO_2^- \rightarrow NO \rightarrow N_2O \rightarrow N_2$) (Hutchinson and Mosier, 1979). Unlike assimilatory NO_3^- reduction, this reaction will occur only under anaerobic conditions and the dissimilatory nitrate reductase is not inhibited by NH_4^+.

2.3.3.2.3 Dissimilatory metal-reducing bacteria. The complete oxidation of organic compounds coupled to the utilization of metals as terminal electron acceptors is a relatively new bacterial process (for example, Fe^{3+} reduction and Mn^{6+} reduction) (Lovley and Phillips, 1988; Myers and Nealson, 1988). These biogeochemical processes were mistakenly attributed to chemical reactions until these pure culture experiments validated the role of bac-

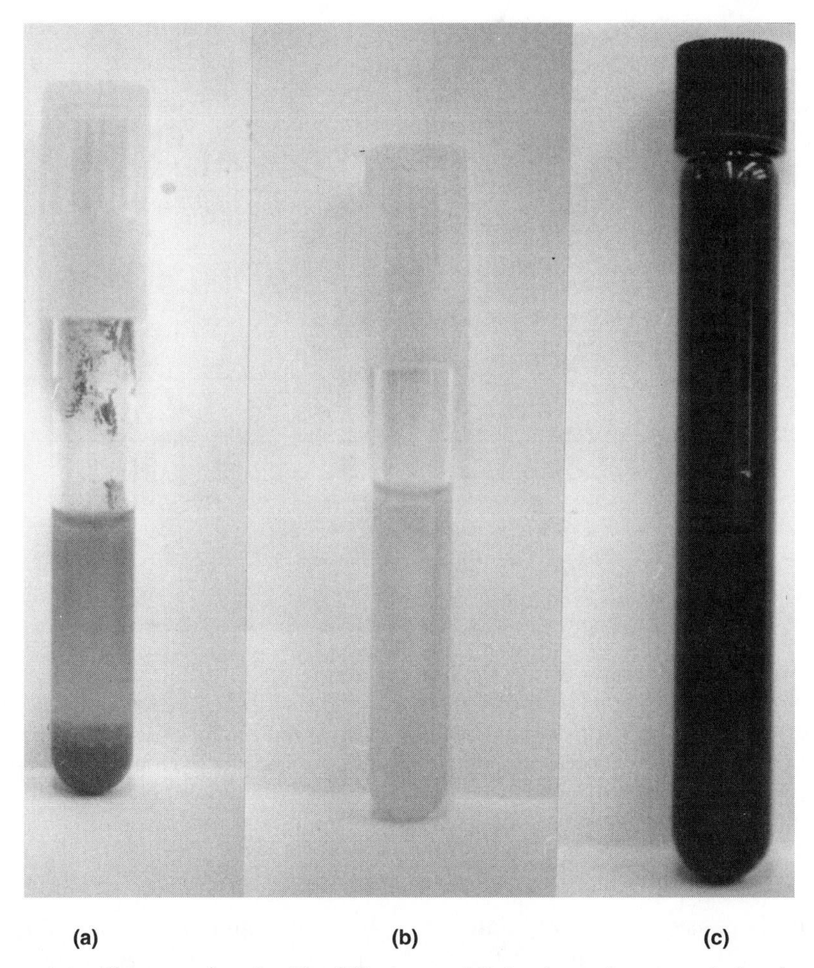

Figure 2-3 Photographs of acidophilic iron-oxidizing bacteria grown in broth media: (a) nitrate-reducing bacteria and (b) sulfate-reducing bacteria. (a) The presence of iron-oxidizing bacteria is reflected in the zone of iron oxyhydroxide and iron sulfate precipitates occurring around the medium–atmosphere interface (see arrowhead). These precipitates are rust-colored, which is diagnostic of iron-oxidizing bacteria. (b) Positive bacterial growth resulting in the presumptive reduction of nitrate needs to be confirmed using the nitrite assay and chemical reduction (control) described in the text. (c) The presence of sulfate-reducing bacteria is indicated by the formation of amorphous iron sulfide (FeS) precipitates in the culture medium. These precipitates are black and are accurately portrayed in the photograph.

teria. Metal-reducing bacteria are now known to contribute to the bioremediation of xenobiotic compounds in anaerobic sediment environments. In addition to iron and manganese reduction, dissimilatory metal reduction has been described for AsO_4^- to AsO_3^- (Laverman et al., 1995),

$Cr(OH)_3^{3+}$ to $Cr(OH)_{3(\bullet)}$ (Ishibashi et al., 1990; Lovley and Phillips, 1994), SeO_4^{2-} to Se° (Oremland et al., 1989), and UO_2^{2+} to $UO_{2(\bullet)}$ (Lovley et al., 1991; Gorby and Lovley, 1992). The formation of metal phosphates through phosphatase activity also represents a novel enzyme-mediated metal precipitation mechanism (Macaskie et al., 1987).

2.3.3.2.4 Dissimilatory sulfate-reducing bacteria. Utilizing SO_4^{2-} as their terminal electron acceptor, sulfate-reducing bacteria (SRB) are the only organisms that can carry out dissimilatory sulfate reduction. Substrates are species specific and include sugars, fatty acids, and H_2. The by-products of their fermentation are H_2S, acetate, CO_2, and H_2O. Dissimilatory sulfate reduction is inhibited by the presence of O_2, nitrate, and ferric iron and is often carbon limited. FeS is a good general indicator of the presence of SRB (Figure 2-3c). The extreme reactivity of HS^- with heavy metals is responsible for the toxicity of SRB activity toward aerobic organisms (plant roots and nematodes in sediments) (Atlas and Bartha, 1993). The HS^- produced reacts with the heavy metals in their cytochrome system. Sulfate reducers can grow over a broad range of environmental conditions. In vitro, SRB do not grow below pH 5 but in acidic (AMD) aquatic environments (pH 3) (Fortin et al., 1995), they may be enriched compared to nonimpacted regions (Gyure et al., 1987).

2.3.4 Archaeobacteria

The archaeobacteria are a group of phylogenetically related bacteria that have been separated from other prokaryotes based on 16S ribosomal ribonucleic acid (rRNA) (Woese, 1981). They are unique in that they are found in extreme environments in terms of anaerobiosis, water activity, and temperature.

2.3.4.1 Methanogenic bacteria

Methanogens are obligate anaerobes which form methane through the reduction of CO_2 to CH_4 (Zeikus, 1977) and fix CO_2 as their carbon source. Alternative substrates may include carbon dioxide, formate, acetate (proximal methyl group), methanol, methylamine, dimethylamine, trimethylamine, and carbon monoxide. They are an extremely important group of bacteria because of their essential function in the last step of the carbon cycle. Methanogens are also important for their contribution to atmospheric methane levels. Global tropospheric methane concentrations average about 1.7 $\mu g \cdot g^{-1}$. This concentration has increased two- to threefold during the past 100 to 200 years, with the present annual increase at 1 to 1.5% (Dacey et al., 1994). The absorption of infrared energy by methane is approximately five times more efficient than that by carbon dioxide. Increased methane flux may be responsible, in part, for the current trend in global warming.

Methanogens are difficult to grow, as they require the anaerobic manipulation of samples and culture media. They are obligate anaerobes which are highly sensitive to oxygen and therefore are located deep within the sediment. In addition, their growth requires several weeks to months and can only be confirmed through methane production. For these reasons, we recommend that samples which require the characterization of methanogen populations be analyzed directly for methane.

Methanogenesis is inhibited by the following (anaerobic) terminal electron acceptors: nitrate, ferric iron (Mn^{4+}), and sulfate. These electron acceptors enable the growth of more energetically efficient microorganisms: dissimilatory nitrate reducers, dissimilatory iron (manganese) reducers, and dissimilatory sulfate reducers, respectively.

2.3.4.2 *Extremely halophilic bacteria*

Halophilic and halotolerant bacteria are found associated with hypersaline environments (e.g., Great Salt Lake, Utah) where the rate of evapotranspiration exceeds the rate of precipitation. These bacteria are characterized by their growth at extremely low water activity, requirement for high concentrations of NaCl (i.e., greater than 2 *M*), and ability to derive energy from sunlight using the protein bacteriorhodopsin as a proton pump (Orem, 1988).

2.3.4.3 *Hyperthermophilic bacteria*

Natural habitats for thermophiles are provided by regions in close proximity to volcanic activity (Atlas and Bartha, 1993). These bacteria typically utilize anaerobic chemolithotrophic metabolic processes (oxidizing hydrogen as an energy source, the reduction of elemental sulfur to sulfide). The occurrence of aerobic, thermophilic (acidophilic, optimum pH between 2 and 3) bacteria such as *Sulfolobus* spp. (sulfide or elemental sulfur oxidation to sulfate) is also restricted to unique hydrothermal environments (e.g., hot sulfur springs).

2.4 *General description of microbiological methods*

Processing sediment samples for microbiological analysis requires the collection of a representative sample that reflects the diversity and density of organisms in the entire ecosystem. As most sediment samples contain anaerobic populations, care must be taken to exclude oxygen from the samples after collection, during storage, and prior to processing of the core or grab samples. For a complete description of sediment sampling and processing, refer to Mudroch and MacKnight (1994) and Mudroch and Azcue (1995).

Most sediment samples will contain relatively large populations of bacteria compared to those from the overlying water column. Therefore, dilution of samples will usually be required prior to plating or dispensing into liquid broth. For descriptions of basic microbiological methodology such as the aseptic technique, serial dilutions, spread plating, and the most probable number (MPN) method, refer to a general microbiology text (e.g., *Biology of Microorganisms* by Brock et al. [1994]).

For MPN determinations, the total number of tubes which produce a positive growth indicator of the desired bacterium is scored. The MPN test will generate a number from each set of five tubes (for example, 5/5, 4/5, 2/5, 0/5, 0/5). To calculate the MPN of organisms in the original sample, use the highest dilution with the greatest number of positive tubes to select p_1, and let p_2 and p_3 represent the number of positive tubes in the next two higher dilutions. Then find the row of numbers in Table 2-3 in which p_1 and p_2 correspond to the values observed experimentally. Follow that row of numbers across the table to the column headed by the observed value of p_3. The figure at the point of the intersection is the MPN of organisms in the quantity of the original sample represented in the inoculum added in the p_2 dilution. Multiply this figure by the appropriate dilution factor to obtain the MPN for the original sample. In the example above, the significant number to be used in conjunction with Table 2-3 to determine the MPN would be 542.

Standard anaerobic techniques include passing gases (e.g., 80:20 N_2:CO_2) through hot (reduced) copper filings to remove traces of oxygen and performing all manipulations (transfers and samplings) under a stream of oxygen-free gas. For a description of standard anaerobic techniques, see Bryant (1972) and Miller and Wolin (1974), who describe a serum-bottle technique for cultivating obligate anaerobes. Both of these references are recommended for review prior to initiating the study of anaerobes.

2.5 *Descriptions of bacterial growth media*

2.5.1 *Chemolithotrophic eubacteria*

2.5.1.1 *Methylotrophic/methanotrophic bacteria*

A nitrate minimal medium and culture conditions for methylotrophs are described by Roslev and King (1994). The medium contains (g/L): KNO_3 (1.0), Na_2HPO_4 (0.9), KH_2PO_4 (0.5), $MgSO_4$ (0.25), Na_2SO_4 (0.1), and $CaCl_2$ (0.1). A trace element solution (10 mL/L) added prior to autoclaving contains (g/L): $ZnCl_2$ (0.025), $CuCl_2$ (0.025), $MnCl_2$ (0.025), NaBr (0.01), Na_2MoO_2 (0.01), KI (0.02), H_3BO_3 (0.01), $CoCl_2$ (0.01), and $NiCl_2$ (0.01). After autoclaving, a filter-sterilized iron solution (1 *M* $FeSO_4$ in 1 *M* HCl) is added (20 mL/L) to the completed medium, which can then be dispensed, aseptically, into sterile Wheaton-brand serum-type culture vessels and capped with butyl rubber stoppers and aluminum seals. Cells can be grown in batch culture with an initial atmosphere of 30% (vol/vol) methane and 70% (vol/vol) air. The fluid-to-head-space ratio for the completed medium should be 1:9.

2.5.2 *Chemoorganotrophic eubacteria*

2.5.2.1 *Aerobic heterotrophic bacteria*

The R2-A growth medium consists of (g/L): yeast extract (0.5), Proteose peptone 3 (0.5), Casamino acids (0.5), dextrose (0.5), soluble starch (0.5),

Table 2-3 Most Probable Numbers for Use with Tenfold Serial Dilutions and Five Tubes per Dilution

		MPN for indicated values by p_3					
p_1	p_2	0	1	2	3	4	5
0	0	—	0.018	0.036	0.054	0.072	0.09
0	1	0.018	0.036	0.055	0.073	0.091	0.11
0	2	0.037	0.055	0.074	0.092	0.11	0.13
0	3	0.056	0.074	0.093	0.11	0.13	0.15
0	4	0.075	0.094	0.11	0.13	0.15	0.17
0	5	0.094	0.11	0.13	0.15	0.17	0.19
1	0	0.020	0.040	0.060	0.081	0.10	0.12
1	1	0.040	0.061	0.081	0.10	0.12	0.14
1	2	0.061	0.082	0.10	0.12	0.15	0.17
1	3	0.083	0.10	0.13	0.15	0.17	0.19
1	4	0.11	0.13	0.15	0.17	0.19	0.22
1	5	0.13	0.15	0.17	0.19	0.22	0.24
2	0	0.045	0.068	0.091	0.12	0.14	0.16
2	1	0.068	0.092	0.12	0.14	0.17	0.19
2	2	0 093	0.12	0.14	0.17	0.19	0.22
2	3	0.12	0.14	0.17	0.20	0.22	0.25
2	4	0.15	0.17	0.20	0.23	0.25	0.28
2	5	0.17	0.20	0.23	0.26	0.29	0.32
3	0	0.078	0.11	0.13	0.16	0.20	0.23
3	1	0 11	0.14	0.17	0.20	0.23	0.27
3	2	0.14	0.17	0.20	0.24	0.27	0.31
3	3	0.17	0.21	0.24	0.28	0.31	0.40
3	4	0.21	0.24	0.28	0.32	0.36	0.40
3	5	0.25	0.29	0.32	0.37	0.41	0.45
4	0	0.13	0.17	0.21	0.25	0.30	0.36
4	1	0.17	0.21	0.26	0.31	0.36	0.42
4	2	0.22	0.26	0.32	0.38	0.44	0.50
4	3	0.27	0.33	0.39	0.45	0.52	0.59
4	4	0.34	0.40	0.47	0.54	0.62	0.69
4	5	0.41	0.48	0.56	0.64	0.72	0.81
5	0	0.23	0.31	0.43	0.58	0.76	0.95
5	1	0.33	0.46	0.64	0.84	1.1	1.3
5	2	0.49	0.70	0.95	1.2	1.5	1.8
5	3	0.79	1.1	1.4	1.8	2.1	2.5
5	4	1.3	1.7	2.2	2.8	3.5	4.3
5	5	2.4	3.5	5.4	9.2	16	—

After Cochran, 1950.

sodium pyruvate (0.3), Na_2HPO_4 (0.3), $MgSO_4$ (0.5), and agar (15) (Amy et al., 1992). Medium preparation consists of combining the appropriate ingredients and autoclaving prior to dispensing into petri dishes. Growth of aerobic heterotrophs is accomplished using serial dilutions in $0.85\%_{(aq)}$ (wt/vol) saline and standard spread-plating methods (Figure 2-2).

2.5.2.2 Anaerobic heterotrophic bacteria

2.5.2.2.1 Dissimilatory nitrate-reducing bacteria. The growth of nitrate reducers can be accomplished using a relatively simple, commercially available growth medium. Nutrient broth (8 g/L, Difco®) can be supplemented with KNO_3 (1 g/L) or the complete commercial medium (Nitrate broth, Difco®) can be used directly. This medium is dispensed in 13 × 100 mm test tubes that contain Durham tubes (see Figure 2-3b). These tubes are inoculated and incubated for 48 hr. One milliliter of the medium is removed to a clean test tube to assay for the presence of nitrite. To this 1-mL sample, 100 μL of 0.8% sulfanilic acid (by weight) in 5 *M* acetic acid is added and vortexed; then 100 μL of 0.6% *N,N*-dimethyl-1-naphthylamine (by weight) in 5 *M* acetic acid is added and vortexed (Wallace and Neave, 1927). When nitrites are mixed with sulfanilic acid, a diazotized sulfanilic acid forms. This reacts with the dimethylnaphthylamine to produce a red azo dye, usually within 10 min (a positive test). If this nitrite test is negative, a very small amount of zinc (covering just the tip of a toothpick) must be added. If there is residual nitrate in the sample, the zinc will chemically reduce it to nitrite and the red azo dye will form. This is a negative test for nitrate reduction. If the sample remains colorless after the addition of zinc, the bacteria have reduced the nitrate beyond nitrite (there may be a positive gas test). An uninoculated control medium should also be tested.

2.5.2.2.2 Dissimilatory iron- and/or manganese-reducing bacteria. Caccavo et al. (1994) described the following culture medium for dissimilatory iron-reducing bacteria. The basal medium contains the following (g/L): $NaHCO_3$ (2.5), NH_4Cl (1.5), KH_2PO_4 (0.6), KCl (0.1), vitamin mix (10 mL), and trace minerals (10 mL). Sodium acetate (10 m*M*) was added as the electron donor and soluble ferric PP_i (3 g/L) was added as the electron acceptor.

The vitamin mix is described by Staley (1968) and consists of (mg/L): *p*-aminobenzoic acid (5.0), calcium pantothenate (5.0), nicotinamide (5.0), thiamine · HCl (5.0), riboflavin (5.0), biotin (2.0), folic acid (2.0), pyridoxine · HCl (2.0), and B_{12} (0.1). The trace mineral solution is described by Balch et al. (1979) and consists of (g/L): nitrilotriacetic acid (1.5), $MgSO_4 \cdot 7H_2O$ (3), $MnSO_4 \cdot 2H_2O$ (0.5), NaCl (1), $FeSO_4 \cdot 7H_2O$ (0.1), $CoCl_2$ (0.1), $CaCl_2 \cdot 2H_2O$ (0.1), $ZnSO_4$ (0.1), Cu $SO_4 \cdot 5H_2O$ (0.01), $AlK(SO_4)_2$ (0.01), and H_3BO_3 (0.01). To prepare the mineral solution, the nitrilotriacetic acid is dissolved with KOH to pH 6.5 and the other minerals are then added.

Amorphous ferric oxyhydroxides are formed by neutralizing a 0.4 *M* solution of $FeCl_3$ to a pH of 7 with NaOH (Lovley and Phillips, 1986). The precipitates are then washed several times with distilled water to remove the unreacted Cl. The final medium should contain approximately 250 m*M* of ferric iron. For manganese-reducing bacteria, MnO_2 is prepared by slowly adding a solution of $MnCl_2$ (30 m*M*) to a basic solution of $KMnO_4$ (20 m*M*) on a magnetic stirrer and washing as described above. Either terminal electron acceptor may be incorporated into the above culture medium.

All solutions must be boiled and cooled under a constant stream of 80% N_2–20% CO_2, dispensed into serum-type culture vessels (i.e., 18 × 150 mm anaerobic pressure tubes [Bellco Glass, Inc., Vineland, New Jersey]) and capped with butyl rubber stoppers and aluminum seals. Saline (0.85% [wt/vol] NaCl) which has been boiled under oxygen-free gas as described above should be used for the dilution series. The MPN method should be used for the enumeration of dissimilatory iron- and/or manganese-reducing bacteria. Positive growth of these bacteria is indicated by solubilization/removal of the rust-colored (or brown) precipitates from the culture vessels.

2.5.2.2.3 Dissimilatory sulfate-reducing bacteria. Growth of most SRB can be accomplished using the following culture medium containing (g/L): Bacto® Tryptone (10.0), $MgSO_4 \cdot 7H_2O$ (2.0), $FeSO_4 \cdot 7H_2O$ (0.5), Na_2SO_3 (0.5), and 60% sodium lactate (6.0 mL/L). The final solution is adjusted to pH 7.5 with 2 *N* NaOH. This medium must be filter sterilized (to prevent iron oxidation and precipitation) and dispensed into sterilized screw-cap test tubes. The volume of dispensed medium should equal 90% of the capacity of the screw-cap test tube minus 1 mL (the sample volume). A reducing agent supplement (RAS; ascorbic acid [7.5 g/L] and thioglycollic acid [7.5 g/L], adjusted to pH 7.5 with 2 *N* NaOH) is added to the medium at 10% (vol/vol of the screw-cap test tube) to maintain reducing conditions. Growth of SRB can be accomplished in screw-cap test tubes through the use of a reducing agent. Anaerobic saline (8 mL 0.85% [wt/vol] NaCl plus 1 mL RAS) must be used for serial dilutions. Quantitative measurement of SRB (Figure 2-3c) populations using basic laboratory equipment requires the MPN method.

2.6 Discussion, recommendations, and conclusions

Understanding the structure and functioning of ecosystems requires quantitative information about (1) numbers of organisms, (2) biomass of populations, (3) rates of activities, (4) rates of growth and death, and (5) cycling and transfer of materials within ecosystems (Atlas and Bartha, 1993).

In the natural environment, bacteria grow as mixed populations which compete with each other for the available inorganic and organic nutrients. The interactions between these bacteria in vivo cannot be exactly modeled in vitro. When enumerating microorganisms from the natural environment, it is typically assumed that these isolated organisms were metabolically active in that environment. Also, in vitro measurements of bacterial populations are inherently underestimated because total bacterial cell counts from natural samples are typically on the order of 100 times greater than in vitro viable cell counts. Certainly, a proportion of the total cell count represents dead bacterial cells. However, the difference between the total cell count and the viable cell count in a sample is usually attributed to nonculturable bacteria.

There is no single culture medium capable of supporting the entire range of bacterial physiological groups. All media constituents and the physico-chemical conditions of growth are selective. Therefore, care must be taken in interpreting in vitro measurements of natural samples. For this reason, it is important to focus any study on the physiological groups which are known or suspected to be important to the research objective(s) at hand. Today, most sediment studies focus on the remediation of environmental contaminants.

The aerobic degradation of organic pollutants is often achieved by microorganisms which utilize these compounds as their carbon source (for example, the mineralization of the herbicide atrazine by a *Pseudomonas* sp. [Yanze-Kontchou and Gschwind, 1994] and the degradation of 1,1,1-trichloro-2,2-bis(4-chlorophenyl)ethane [DDT] by *Alcaligenes eutrophus* A5 [Nadeau et al., 1994]).

The anaerobic degradation of aromatic compounds is known to occur under nitrate-reducing (Kuhn et al., 1988; Al Bashir et al., 1990; Dolfing et al., 1990; Evans et al., 1991; Hutchins, 1991; Schocher et al., 1991; Barbaro et al., 1992; Flyvbjerg et al., 1993; Fries et al., 1994), ferric iron-reducing (Lovley et al., 1989; Lovley and Lonergan, 1990; Kazumi et al., 1995), sulfate-reducing (Haag et al., 1991; Beller et al., 1992; Edwards and Grbic-Galic, 1992; Flyvbjerg et al., 1993; Rabus et al., 1993; Haagblom and Young, 1995; Lovley et al., 1995), and methanogenic conditions (Vogel and Grbic-Galic, 1986; Wilson et al., 1986; Grbic-Galic and Vogel, 1987; Edwards and Grbic-Galic, 1994). Anaerobic conditions also provide a favorable environment for reductive dehalogenation under nitrate-reducing, iron-reducing (Kazumi et al., 1995), and sulfate-reducing conditions (Sonier et al., 1994; Haagblom and Young, 1995). Because of the microenvironments experienced by bacteria, reductive dechlorination of trichloroethylene and tetrachloroethylene has also been measured under (macroscopic) aerobic conditions in a sediment column (Enzien et al., 1994).

The biogeochemistry of metal wastes as it relates to the distribution and transformation of toxic heavy metals also deserves attention in studies of sediment processes. In addition to the characterization of dissimilatory metal reduction, bacteria are capable of transforming metals by a variety of other mechanisms, for example, mercury methylation by *Desulfovibrio desulfuricans* LS, forming volatile methyl mercury (Choi et al., 1994), and selenium methylation, forming volatile dimethyl selenide (Thompson-Eagle and Frankenburger, 1991). Bacteria, as a group, also act as nucleating agents of metals, forming mineralized microfossils in aerobic and anaerobic environments (Ferris et al., 1987).

It is our hope that these general guidelines for the bacteriological examination of sediments will be beneficial to studies of biogeochemical processes in sediments. However, a cautionary note must be included. The recovery of bacteria from any environment is a measurement of the metabolic potential of the sample and does not necessarily reflect the dominant metabolic processes occurring in the sediment at the time of sampling. Therefore, the enumeration of bacterial populations in vitro cannot alone provide conclusive evidence of the biogeochemical processes occurring within a sediment environment. However, in concert with chemical analyses, bacteriological data should enable a more conclusive interpretation of actual and potential processes occurring within sediments.

References

Adams, L.F. and Ghiorse, W.C., Physiology and ultrastructure of *Leptothrix discophora* SS-1, *Arch. Microbiol.*, 145, 126, 1986.

Adams, L.F. and Ghiorse, W.C., Characterization of extracellular Mn^{2+}-oxidizing activity and isolation of an Mn^{2+}-oxidizing protein from *Leptothrix discophora* SS-1, *J. Bacteriol.*, 169, 1279, 1987.

Al Bashir, B., Cseh, T., Leduc, R., and Samson, R., Effect of soil/contaminant interactions on the biodegradation of naphthalene in flooded soils under denitrifying conditions, *Appl. Microbiol. Biotechnol.*, 34, 414, 1990.

Amy, P.S., Haldeman, D.L., Ringelberg, D., Hall, D.H., and Russell, C., Comparison of identification systems for classification of bacteria isolated from water and endolithic habitats within the deep subsurface, *Appl. Environ. Microbiol.*, 58, 3367, 1992.

Atlas, R.M. and Bartha, R., *Microbial Ecology: Fundamentals and Applications*, 3rd ed., Benjamin/Cummings, Redwood, CA, 1993, 563.

Balch, W.E., Fox, G.E., Magrum, L.J., Woese, C.R., and Wolfe, R.S., Methanogens: reevaluation of a unique biological group, *Microbiol. Rev.*, 43, 260, 1979.

Barbaro, J.R., Barker, J.F., Lemon, L.A., and Mayfield, C.I., Biotransformation of BTEX under anaerobic denitrifying conditions: field and laboratory observations, *J. Contam. Hydrol.*, 11, 245, 1992.

Beller, H.R., Grbic-Galic, D., and Reinhard, M., Microbial degradation of toluene under sulfate-reducing conditions and the influence of iron on the process, *Appl. Environ. Microbiol.*, 58, 786, 1992.

Beveridge, T.J., Ultrastructure, chemistry, and function of the bacterial cell wall, *Int. Rev. Cytol.*, 72, 229, 1981.

Boogerd, F.C. and de Vrind, J.P.M., Manganese oxidation by *Leptothrix discophora*, *J. Bacteriol.*, 169, 489, 1987.

Brock, T.D., Madigan, M.T., Martinko, J.M., and Parker, J., Eds., *Biology of Microorganisms*, 7th ed., Prentice-Hall, New York, 1994, 909.

Bryant, M.P., Commentary on the Hungate technique for culture of anaerobic bacteria, *Am. J. Clin. Nutr.*, 25, 1324, 1972.

Caccavo, F., Lonergan, D.J., Lovley, D.R., Davis, M., Stolz, J.F., and McInerney, M.J., *Geobacter sulfurreducens* sp. nov., a hydrogen- and acetate-oxidizing dissimilatory metal-reducing microorganism, *Appl. Environ. Microbiol.*, 60, 3752, 1994.

Caldwell, D.R., *Microbial Physiology and Metabolism*, Wm. C. Brown, Dubuque, IA, 1995, 353.

Chakrabarti, B.K. and Banerjee, P.C., Surface hydrophobicity of acidophilic heterotrophic bacterial cells in relation to their adhesion on minerals, *Can. J. Microbiol.*, 37, 692, 1991.

Choi, S.C., Chase, T., and Bartha, R., Metabolic pathways leading to mercury methylation in *Desulfovibrio desulfuricans* LS, *Appl. Environ. Microbiol.*, 60, 4072, 1994.

Cochran, W.G., Estimation of bacterial densities by means of the "most probable number," *Biometrics*, 6, 105, 1950.

Corstjens, P.L.A.M., de Vrind, J.P.M., Westbroek, P., and de Vrind–de Jong, E.W., Enzymatic iron oxidation by *Leptothrix discophora*: identification of an iron-oxidizing protein, *Appl. Environ. Microbiol.*, 58, 450, 1992.

Crerar, D.A., Knox, G.W., and Means, J.L., Biogeochemistry of bog iron in the New Jersey pine barrens, *Chem. Geol.*, 24, 111, 1979.

Dacey, J.W.H., Drake, B.G., and Klug, M.J., Stimulation of methane emission by carbon dioxide enrichment of marsh vegetation, *Nature*, 370, 47, 1994.

Dolfing, J., Zeyer, J., Binder-Eicher, P., and Schwarzenbach, R.P., Isolation and characterization of a bacterium that mineralizes toluene in the absence of molecular oxygen, *Arch. Microbiol.*, 154, 336, 1990.

Edwards, E.A. and Grbic-Galic, D., Complete mineralization of benzene by aquifer microorganisms under strictly anaerobic conditions, *Appl. Environ. Microbiol.*, 58, 2663, 1992.

Edwards, E.A. and Grbic-Galic, D., Anaerobic degradation of toluene and *o*-xylene by a methanogenic consortium, *Appl. Environ. Microbiol.*, 60, 313, 1994.

Emerson, D. and Ghiorse, W.C., Isolation, cultural maintenance, and taxonomy of a sheath-forming strain of *Leptothrix discophora* and characterization of manganese-oxidizing activity associated with the sheath, *Appl. Environ. Microbiol.*, 58, 4001, 1992.

Emerson, D. and Revsbech, N.P., Investigation of an iron-oxidizing microbial mat community located near Aarhus, Denmark: field studies, *Appl. Environ. Microbiol.*, 60, 4022, 1994a.

Emerson, D. and Revsbech, N.P., Investigation of an iron-oxidizing microbial mat community located near Aarhus, Denmark: laboratory studies, *Appl. Environ. Microbiol.*, 60, 4032, 1994b.

Enzien, M.V., Picardal, F., Hazen, T.C., Arnold, R.G., and Flieermans, C.B., Reductive dechlorination of trichloroethylene and tetrachloroethylene under aerobic conditions in a sediment column, *Appl. Environ. Microbiol.*, 60, 2200, 1994.

Evans, P.J., Mang, D.T., Kim, K.S., and Young, L.Y., Anaerobic degradation of toluene by a denitrifying bacterium, *Appl. Environ. Microbiol.*, 57, 1139, 1991.

Fenchel, T.M. and Jorgensen, B.B., Detritus food chains of aquatic ecosystems: the role of bacteria, *Adv. Microbial Ecol.*, 1, 1, 1977.

Ferris, F.G., Fyfe, W.S., and Beveridge, T.J., Bacteria as nucleation sites for authigenic minerals in a metal-contaminated lake sediment, *Chem. Geol.*, 63, 225, 1987.

Flyvbjerg, J., Arivn, E., Jensen, B.K., and Olsen, S.K., Microbial degradation of phenols and aromatic hydrocarbons in creosote-contaminated groundwater under nitrate-reducing conditions, *J. Contam. Hydrol.*, 12, 133, 1993.

Focht, D.D. and Verstraete, W., Biochemical ecology of nitrification and denitrification, *Adv. Microbial Ecol.*, 1, 135, 1977.

Fogel, M.M., Taddeo, A.R., and Fogel, S., Biodegradation of chlorinated ethanes by a methane-utilizing mixed culture, *Appl. Environ. Microbiol.*, 51, 720, 1986.

Fortin, D., Davis, B., Southam, G., and Beveridge, T.J., Biogeochemical phenomena induced by bacteria within sulfidic mine tailings, *J. Ind. Microbiol.*, 14, 178, 1995.

Fries, M.R., Zhou, J., Chee-Sanford, J., and Tiedje, J.M., Isolation, characterization, and distribution of denitrifying toluene degraders from a variety of habitats, *Appl. Environ. Microbiol.*, 60, 2802, 1994.

Fukui, M. and Takii, S., Colony formation of free-living and particle-associated sulfate-reducing bacteria, *FEMS Microbiol. Ecol.*, 73, 85, 1990.

Gadd, G.M., Microbial formation and transformation of organometallic and organometalloid compounds, *FEMS Microbiol. Rev.*, 11, 297, 1993.

Goldman, M. and Wilson, D.A., Growth of *Sporosarcina ureae* in defined media, *FEMS Microbiol. Lett.*, 2, 113, 1977.

Goodwin, S. and Zeikus, J.G., Ecophysiological adaptations of anaerobic bacteria to low pH: analysis of anaerobic digestion in acidic bog sediments, *Appl. Environ. Microbiol.*, 53, 57, 1987.

Gorby, Y.A. and Lovley, D.A., Enzymatic uranium precipitation, *Environ. Sci. Technol.*, 26, 205, 1992.

Grbic-Galic, D. and Vogel, T., Transformation of toluene and benzene by mixed methanogenic cultures, *Appl. Environ. Microbiol.*, 53, 254, 1987.

Gyure, R.A., Konopka, A., Brooks, A., and Doemel, W., Algal and bacterial activities in acidic (pH 3) strip-mine lakes, *Appl. Environ. Microbiol.*, 53, 2069, 1987.

Haag, F., Reinhard, M., and McCarty, P.L., Degradation of toluene and *p*-xylene in anaerobic microcosms: evidence for sulfate, a terminal electron acceptor, *Environ. Toxicol. Chem.*, 10, 1379, 1991.

Haagblom, M.M. and Young, L.Y., Anaerobic degradation of halogenated phenols by sulfate-reducing consortia, *Appl. Environ. Microbiol.*, 61, 1546, 1995.

Holt, J.G., *Bergey's Manual of Systematic Bacteriology*, Murray, R.G.E., Brenner, D.J., Bryant, M.P., Holt, J.G., Krieg, N.R., Moulder, J.W., Pfennig, N., Sneath, P.H.A., and Staley, J.T., Eds., Williams and Wilkins, Baltimore, 1984.

Hubel, H., Die ^{14}C methode zur Bestimmung der primarproduktion des Phytoplanktons, *Limnologica*, 4, 267, 1966.

Hutchins, S.R., Optimizing BTEX biodegradation under denitrifying conditions, *Environ. Toxicol. Chem.*, 10, 1437, 1991.

Hutchinson, G.L. and Mosier, A.R., Nitrous oxide emissions from an irrigated cornfield, *Science*, 205, 1125, 1979.

Ishibashi, Y., Cervantes, C., and Silver, S., Chromium reduction in *Pseudomonas putida*, *Appl. Environ. Microbiol.*, 56, 2268, 1990.

Jensen, K., Sloth, N.P., Risgaard-Petersen, N., Rysgaard, S., and Revsbech, N.P., Estimation of nitrification and denitrification from microprofiles of oxygen and nitrate in model sediment systems, *Appl. Environ. Microbiol.*, 60, 2094, 1994.

Kazumi, J., Haagblom, M.M., and Young, L.Y., Degradation of monochlorinated and nonchlorinated aromatic compounds under iron-reducing conditions, *Appl. Environ. Microbiol.*, 61, 4069, 1995.

Knaebel, D.B., Federle, T.W., McAvoy, D.C., and Vestal, J.R., Effect of mineral and organic soil constituents on microbial mineralization of organic compounds in a natural soil, *Appl. Environ. Microbiol.*, 60, 4500, 1994.

Konhauser, K.O., Schultze-Lam, S., Ferris, F.G., Fyfe, W.S., Longstaff, F.J., and Beveridge, T.J., Mineral precipitation by epilithic biofilms in the Speed River, Ontario, Canada, *Appl. Environ. Microbiol.*, 60, 549, 1994.

Korber, D.R., Lawrence, J.R., Lappin-Scott, H.M., and Costerton, J.W., Growth of microorganisms on surfaces, in *Microbial Biofilms*, Lappin-Scott, H.M. and Costerton, J.W., Eds., Cambridge University Press, New York, 1995, 15.

Kuhn, E.P., Zeyer, J., Eicher, P., and Schwarzenbach, R.P., Anaerobic degradation of alkylated benzenes in denitrifying laboratory aquifer columns, *Appl. Environ. Microbiol.*, 54, 490, 1988.

Laverman, A.M., Blum, J.S., Schaefer, J.K., Phillips, E.J.P., Lovley, D.R., and Oremland, R.S., Growth of strain SES-3 with arsenate and other diverse electron acceptors, *Appl. Environ. Microbiol.*, 61, 3556, 1995.

Little, C.D., Palumbo, A.V., and Herbes, S.E., Trichloroethylene biodegradation by a methane-oxidizing bacterium, *Appl. Environ. Microbiol.*, 54, 951, 1988.

Ljungdahl, L.G. and Eriksson, K.E., Ecology of microbial cellulose degradation, *Adv. Microbial Ecol.*, 8, 237, 1985.

Lovley, D.R. and Klug, M.J., Sulfate reducers can outcompete methanogens at freshwater sulfate concentrations, *Appl. Environ. Microbiol.*, 45, 187, 1983.

Lovley, D.R. and Lonergan, D.J., Anaerobic oxidation of toluene, phenol, and *p*-cresol by the dissimilatory iron-reducing organism GS-15, *Appl. Environ. Microbiol.*, 56, 1858, 1990.

Lovley, D.R. and Phillips, E.J.P., Organic matter mineralization with the reduction of ferric iron in anaerobic sediments, *Appl. Environ. Microbiol.*, 51, 683, 1986.

Lovley, D.R. and Phillips, E.J.P., Novel mode of microbial energy metabolism: organic carbon oxidation coupled to dissimilatory reduction of iron or manganese, *Appl. Environ. Microbiol.*, 54, 1472, 1988.

Lovley, D.R. and Phillips, E.J.P., Reduction of chromate by *Desulfovibrio vulgaris* and its c_3 cytochrome, *Appl. Environ. Microbiol.*, 60, 726, 1994.

Lovley, D.R., Baedecker, M.J., Lonergan, D.J., Cozzarelli, L.M., Phillips, E.J.P., and Siegel, D.I., Oxidation of aromatic contaminants coupled to microbial iron reduction, *Nature*, 339, 297, 1989.

Lovley, D.R., Phillips, E.J.P., Gorby, Y.A., and Landa, E.R., Microbial reduction of uranium, *Nature*, 350, 413, 1991.

Lovley, D.R., Coates, J.D., Woodward, J.C., and Phillips, E.J.P., Benzene oxidation coupled to sulfate reduction, *Appl. Environ. Microbiol.*, 61, 953, 1995.

Macaskie, L.E., Dean, A.C.R., Cheetham, A.K., Jakeman, R.J.B., and Skarnulis, A.J., Cadmium accumulation by a *Citrobacter* sp.: the chemical nature of the accumulated metal precipitate and its location on the bacterial cells, *J. Gen. Microbiol.*, 133, 539, 1987.

Miller, T.L. and Wolin, M.J., A serum bottle modification of the Hungate technique for cultivating obligate anaerobes, *Appl. Environ. Microbiol.*, 27, 985, 1974.

Mills, A.L., Herman, J.S., Hornberger, G.M., and DeJesus, T.H., Effect of solution ionic strength and iron coatings on mineral grains on the sorption of bacterial cells to quartz sand, *Appl. Environ. Microbiol.*, 60, 3300, 1994.

Mudroch, A. and Azcue, J.M., *Manual of Aquatic Sediment Sampling*, Lewis Publishers, Boca Raton, FL, 1995, 219.

Mudroch, A. and MacKnight, S.D., Eds., *Handbook of Techniques for Aquatic Sediments Sampling*, 2nd ed., Lewis Publishers, Boca Raton, FL, 1994, 236.

Myers, C.P. and Nealson, K.H., Bacterial manganese reduction and growth with manganese oxide as the sole electron acceptor, *Science*, 240, 1319, 1988.

Nadeau, L.J., Menn, F.-M., Breen, A., and Sayler, G.S., Aerobic degradation of 1,1,1-trichloro-2,2-bis(4-chlorophenyl)ethane (DDT) by *Alcaligenes eutrophus* A5, *Appl. Environ. Microbiol.*, 60, 51, 1994.

Orem, A., The microbial ecology of the Dead Sea, *Adv. Microbial Ecol.*, 109, 193, 1988.

Oremland, R.S., Hollibaugh, J.T., Maest, A.S., Presser, T.S., Miller, L.S., and Culbertson, C.W., Selenate reduction to elemental selenium by anaerobic bacteria in sediments and culture: biogeochemical significance of a novel sulfate-independent respiration, *Appl. Environ. Microbiol.*, 55, 2333, 1989.

Rabus, R., Nordhaus, R., Ludwig, W., and Widdel, F., Complete oxidation of toluene under strictly anoxic conditions by a new sulfate-reducing bacterium, *Appl. Environ. Microbiol.*, 59, 1444, 1993.

Revsbech, N.P. and Jorgensen, B.B., Microelectrodes: their use in microbial ecology, *Adv. Microbial Ecol.*, 9, 293, 1986.

Rogers, S.R. and Anderson, J.J., Measurement of growth and iron deposition in *Sphaerotilus discophorus*, *J. Bacteriol.*, 126, 257, 1976a.

Rogers, S.R. and Anderson, J.J., Role of iron deposition in *Sphaerotilus discophorus*, *J. Bacteriol.*, 126, 264, 1976b.

Roslev, P. and King, G.M., Survival and recovery of methanotrophic bacteria starved under oxic and anoxic conditions, *Appl. Environ. Microbiol.*, 60, 2602, 1994.

Schocher, R.J., Seyfroed, B., Vazquez, F., and Zeyer, J., Anaerobic degradation of toluene by pure cultures of denitrifying bacteria, *Arch. Microbiol.*, 157, 7, 1991.

Scow, K.M. and Hutson, J., Effect of diffusion and sorption on the kinetics of biodegradation: theoretical considerations, *Soil Sci. Soc. Am. J.*, 56, 119, 1992.

Silverman, M.P. and Lundgren, D.G., Studies on the chemoautotrophic iron bacterium *Ferrobacillus ferrooxidans*. I. An improved medium and a harvesting procedure for securing high cell yields, *J. Bacteriol.*, 77, 642, 1959.

Sonier, D.N., Duran, N.L., and Smith, G.B., Dechlorination of trichlorofluoromethane (CFC-11) by sulfate reducing bacteria from an aquifer contaminated with halogenated aliphatic compounds, *Appl. Environ. Microbiol.*, 60, 4567, 1994.

Soriano, S. and Walker, N., Isolation of ammonium oxidizing autotrophic bacteria, *J. Appl. Bacteriol.*, 31, 493, 1968.

Southam, G., unpublished data, 1995.

Southam, G. and Beveridge, T.J., Enumeration of thiobacilli within pH-neutral and acidic mine tailings and their role in the development of secondary mineral soil, *Appl. Environ. Microbiol.*, 58, 1904, 1992.

Southam, G., Ferris, F.G., and Beveridge, T.J., Mineralized bacterial biofilms in sulphide tailings and in acid mine drainage systems, in *Microbial Biofilms,* Lappin-Scott, H.M. and Costerton, J.W., Eds., Cambridge University Press, New York, 1995, 148.

Staley, J.T., *Prosthecomicrobium* and *Ancalomicrobium*: new prosthecate freshwater bacteria, *J. Bacteriol.*, 95, 1921, 1968.

Stanier, R.Y., Pfennig, N., and Trüper, H., Introduction to the phototrophic prokaryotes, in *The Prokaryotes: A Handbook on Habitats, Isolation and Identification of Bacteria,* Starr, M.P., Stolp, H., Trüper, H., Ballow, A., and Schlegel, H.G., Eds., Springer-Verlag, Berlin, 1981, 197.

Thompson, J.B., personal communication, 1995.

Thompson-Eagle, E.T. and Frankenburger W.T., Jr., Selenium biomethylation in an alkaline, saline environment, *Water Res.*, 25, 231, 1991.

Urrutia, M. and Beveridge, T.J., Mechanism of silicate binding to the bacterial cell wall in *Bacillus subtilis*, *J. Bacteriol.*, 175, 1936, 1993.

Vogel, T. and Grbic-Galic, D., Incorporation of oxygen from water into toluene and benzene during anaerobic fermentative transformation, *Appl. Environ. Microbiol.*, 52, 200, 1986.

Wallace, G.L. and Neave, S.L., The nitrate test as applied to bacterial cultures, *J. Bacteriol.*, 14, 377, 1927.

Wilson, B.H., Smith, G.B., and Reis, J.F., Biotransformations of selected alkylbenzenes and halogenated aliphatic hydrocarbons in methanogenic aquifer material: a microcosm study, *Environ. Sci. Technol.*, 20, 997, 1986.

Woese, C.R., Archaebacteria, *Sci. Am.*, 244, 98, 1981.

Yanze-Kontchou, C. and Gschwind, N., Mineralization of the herbicide atrazine as a carbon source by a *Pseudomonas* strain, *Appl. Environ. Microbiol.*, 60, 4297, 1994.

Zeikus, J.G., The biology of methanogenic bacteria, *Bacteriol. Rev.*, 41, 514, 1977.

Zeikus, J.G., Kerby, R., and Krzycki, J.A., Single carbon chemistry of acetogenic and methanogenic bacteria, *Science*, 227, 1167, 1985.

chapter three

Laboratory methods and criteria for sediment bioassessment

Pilar Rodriguez and Trefor B. Reynoldson

3.1 Introduction

Several decades ago, farsighted individuals suggested that biological systems were an essential part of ecosystem health assessment and could provide a basis for environmental management (Patrick, 1951; Hynes, 1960). A series of bioassessment techniques, along with appropriate decision-making criteria, are necessary to identify the types of stress being exerted, their severity, and the bioavailability of the contaminants present in the ecosystems. This broader view is particularly well expressed at the policy level in the 1987 revision of the Great Lakes Water Quality Agreement between the United States and Canada, which explicitly endorses an ecosystem approach. This is being increasingly accepted in the scientific literature and by government agencies (Bode and Novak, 1995; Southerland and Stribling, 1995; Yoder and Rankin, 1995).

A fundamental distinction can be made between structural (usually field methods) and functional (both field and laboratory methods) approaches to bioassessment. Much of the earliest work was of the structural or taxonomic type, where changes in community structure and indicator species were related to anthropogenic and naturally induced stresses (Hynes, 1960; Warren, 1971; Whitton, 1975). This kind of information has been a basic tool for classification of sites in some countries, for example, Great Britain (Wright et al., 1984), New Zealand and Australia (Faith and Norris, 1989; Norris, 1994), and Canada (Reynoldson et al., 1995), and will be addressed in Chapter 4. However, a remedial action or regulatory change requires cause–effect data that cannot be easily inferred from structural data alone. Functional tests, such as laboratory bioassays, can provide this necessary information, as they are specific and are usually capable of discriminating the root cause

1-56670-343-3/99/$0.00+$.50

of structural changes as well as quantifying dose–response relationships. Functional biological monitoring can be broadly defined as the measurement of any rate process or response of the ecosystem and can utilize both taxonomic and nontaxonomic parameters (Matthews et al., 1982). Such taxonomic tests include measures of species colonization or emigration rates and the rate of reestablishment of equilibrium densities following perturbation (Cairns et al., 1979). The more frequently used nontaxonomic tests include both lethal and sublethal end points in both short-term and chronic (i.e., longer term) exposures that include measures of behavior, biochemical–physiological changes, bioaccumulation, genotoxicity (i.e., mutagenesis, carcinogenesis, teratogenesis), reproductive impairment, and death. While many of these tests can be conducted *in situ*, the need for strict control over exposure conditions means that most bioassays are conducted in the laboratory. In addition, most of these tests are performed using a single species which may not be indigenous to the area under investigation. Therefore, although laboratory tests can provide quantifiable relationships between contaminants and organisms and isolate sediment as a contributing factor, their lack of environmental realism and failure to incorporate ecosystem complexity limit their investigative utility and possibly their ability to establish actual impairment (Monk, 1983). The obvious solution to this dilemma is to use both field and laboratory methods to examine stress.

Long and Chapman (1985) suggested that adequate assessments of sediment quality should involve the following three components: concentrations of toxic chemicals (i.e., bulk sediment chemistry), laboratory measures of biological effect (i.e., toxicity of environmental samples), and field identification of biological impact (i.e., species composition and densities of the resident biota). Together, these three categories of measurement have been termed the sediment quality triad (Chapman, 1990). The three components of this approach may be viewed as independent and sufficient for a sediment quality approach, but they comprise interdependent end points that together allow an integrated and realistic approach to the assessment of sediment quality.

In a recent contribution on sediment toxicity assessment, Chapman et al. (1992) described the various components of an integrated sediment assessment. These include sediment chemistry, sediment toxicity, tissue chemistry, pathology, and community structure. These five areas provide a sufficiently broad description of the general types of tools available for conducting sediment assessment.

In the present chapter, the focus is on laboratory sediment bioassays that provide a means for measuring biological changes caused by contaminated sediments. Data on the bioassays are estimates of the amount of biologically active material which affects the test organisms (Reynoldson and Day, 1993). The chapter is structured to represent a bioassessment survey (Figure 3-1). The first requirement is to define clear objectives, as several types of studies may require the use of laboratory toxicity bioassessment of sediments (Giesy and Hoke, 1989; Hill et al., 1993; American Society for Testing and Materials [ASTM], 1994). The objectives can be stated as follows:

1. Screening for horizontal and vertical gradients of sediment toxicity in an area or region for a general sediment classification
2. Measuring the efficiency of remediation programs
3. Measuring the impacts of dredging and channelization, generating resuspension of sediment into the water column
4. Regulating effluent discharges
5. Assessing the bioavailability and hazards of chemicals and establishing numerical targets for quality standards and criteria of sediment toxicity
6. Bioaccumulation of toxicants in tissues of aquatic organisms for protection of the ecosystem and human health
7. Establishing a causality between the biological community and contaminants in sediments

The second requirement relates to the criteria used in selecting appropriate tests from the large number of available bioassays. At this stage, decisions have to be made on using sediment phase, exposure time, test species, and relevant end points in the bioassessment. The selection of these components will affect the sampling methods, such as the amount of sediment to be collected, sediment depth, number of samples, and the sampling device. The laboratory testing procedures are grouped into the following three categories:

1. Those that measure the adverse response of organisms exposed to sediments (i.e., whole sediments or sediment extracts) against a control sediment
2. Those that measure the effect of a dose or the no-effect level of a chemical substance in dilution or spiked to sediments
3. Those that measure the detoxification of a polluted sediment by dilution

Finally, detailed descriptions of every bioassay method have not been provided, but an attempt has been made to be as comprehensive as possible in describing the source literature.

3.2 *Criteria for the selection of the bioassay*

3.2.1 *Single-species assay versus battery of assays versus microcosm and mesocosm*

A principal concept in modern toxicological bioassessment is that no single bioassay can be expected to be adequate for the detection of the potential

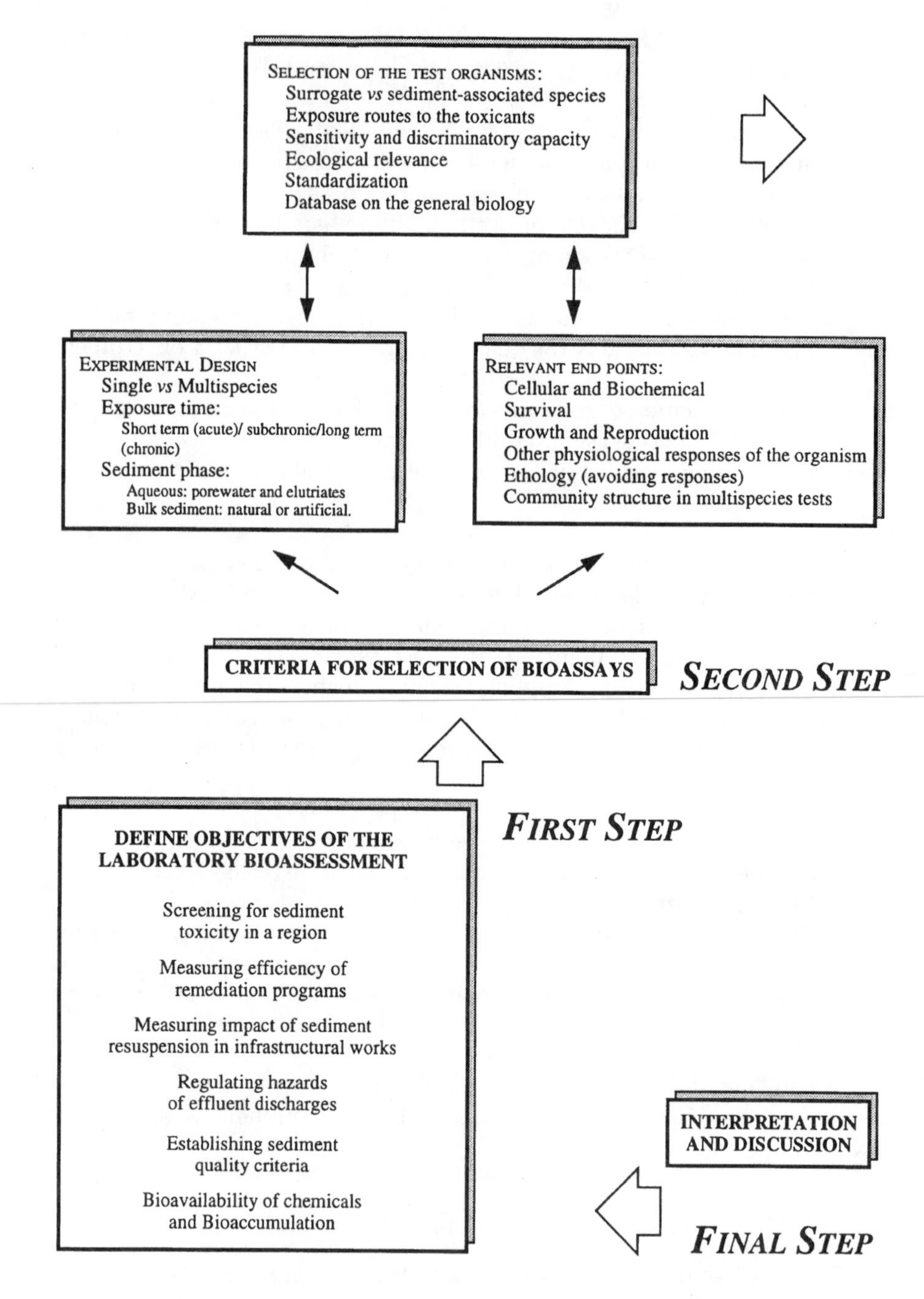

Figure 3-1 General scheme of a sediment bioassessment survey.

THIRD STEP

SAMPLING DECISIONS

SELECTION OF COLLECTION SITES:
- Location
- Number of sites
- Field replicates

SELECTION OF REFERENCE AND CONTROL SEDIMENTS

SAMPLING DEVICES:
Cores, Dredges and Grabs

SAMPLING ISSUES:
- Compositing
- Materials
- Handling
- Storage time and temperature

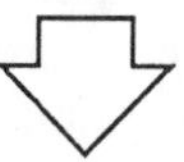

FOURTH STEP

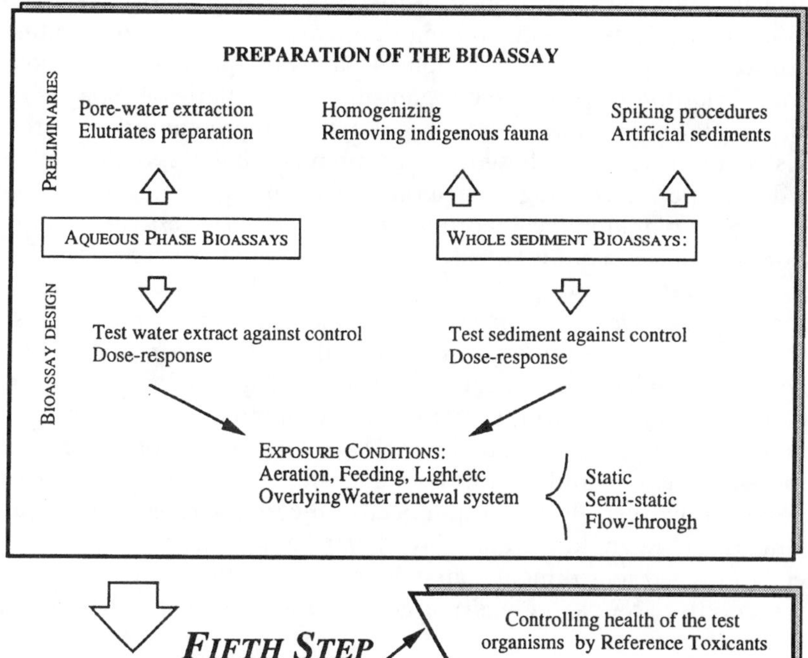

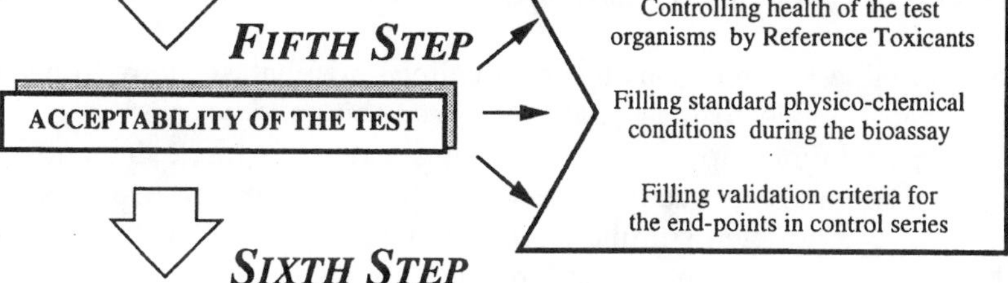

EXPRESSION OF THE RESULTS AND STATISTICAL ANALYSIS

adverse effects of complex mixtures of contaminants (Burton, 1989; Giesy and Hoke, 1989). This is even truer in sediment bioassays, because of the various exposure mechanisms and pathways of toxicants in sediment, such as pore water, association with the solid matrix, or complexation with organic material. These complications have led to the development of batteries of assays that enable interpretation of the effects produced by different exposure routes. Thus, at a minimum, a battery of assays should include organisms that live in contact with the sediment (i.e., epibenthic species), organisms that ingest the sediment (i.e., infaunal detritivores), and possibly organisms living in the water column (i.e., planktonic) or infaunal filtrators of the overlying water. There is no indication of a relationship between exposure route and susceptibility to contaminants (Hill et al., 1993). However, it appears that organisms exposed to different uptake routes, such as contact, adsorption, and ingestion, are more likely to be affected than those exposed to a single uptake pathway, such as the soluble or leachable portion to planktonic organisms. Test batteries of species commonly adopt a strategy to include organisms that represent different trophic levels in the ecosystem, based on the premise that this will provide a broader response coverage: bacteria (Microtox® test); algae (*Chlorella, Selenastrum*); planktonic cladocerans (*Daphnia magna, Ceriodaphnia dubia*); one or several sediment infaunal species, such as oligochaetes (*Tubifex tubifex* or *Lumbriculus variegatus*), amphipods (*Hyalella azteca, Diporeia* sp.), ephemeroptera (*Hexagenia limbata*), or dipterans (*Chironomus tentans, C. riparius*); and a fish (*Pimephales promelas*). Table 3-1 shows several batteries of species being used for sediment bioassessment. Hill et al. (1993) suggested useful selection criteria that can be considered to select test species for a test battery. The multitrophic approach for batteries of tests has been proposed by several authors for more inclusive data on toxicity (Burton et al., 1989). However, others consider benthic invertebrates most appropriate for inclusion (Reynoldson and Day, 1993) because these organisms are directly exposed to sediment contaminants, and furthermore, in comparison with other groups, there is a large database on their response to environmental stress.

Microbial test batteries, including alkaline phosphatase, dehydrogenase, glucosidase and galactosidase, have also been shown to be effective in demonstrating a relationship with the overlying water (Burton, 1989) and have good discriminatory capacity between sites (Burton et al., 1989). Several authors have conducted simultaneous single-species bioassays using recirculating systems and found a relationship among survival, metal content in the sediment, and field distribution of macroinvertebrates (Prater and Anderson, 1977; Malueg et al., 1983, 1984; Nelson et al., 1993). This system allows simultaneous acute and chronic tests in water and sediment phases for toxicity bioassessment.

Multispecies tests at microcosm and mesocosm levels are relevant ecosystem-simulating models used for the evaluation of the fate and effects of

Table 3-1 Batteries of Bioassays for Sediment Bioassessment

Battery of assays for sediment bioassessment	Reference
Microtox®, *Selenastrum capricornutum, Ceriodaphnia dubia, Pimephales promelas*	Ankley et al., 1990
C. dubia, Hyalella azteca, Lumbriculus variegatus, P. promelas	Ankley et al., 1991a
Photobacterium phosphoreum, S. capricornutum, C. dubia, Hexagenia limbata, Pimephales promelas liver lesions in three fishes, mutagenicity, effects in reproduction of two avian species, in vitro bioassays with H4IIE rat hepatoma cells	Ankley et al., 1992b
Hyalella azteca, Chironomus tentans, L. variegatus	Ankley et al., 1993a, 1994b; West et al., 1993
Daphnia magna, Ceriodaphnia dubia, H. azteca, S. capricornutum, enzymatic activity of microbial communities	Burton et al., 1989
Helisoma sp., *L. variegatus, H. azteca*	Di Toro et al., 1992
Microtox®, alga, *D. magna* and *Chironomus tentans*	Giesy and Hoke, 1989
Microtox®, *D. magna, C. tentans*	Giesy et al., 1988b
Microtox®, *D. magna, Hexagenia limbata*	Giesy et al., 1990
Diporeia sp., *C. riparius* and *L. variegatus*	Harkey et al., 1994
Daphnia magna, H. limbata, Lireus fontinalis, P. promelas	Hoke and Prater, 1980
Microtox®, *D. magna, Ceriodaphnia dubia, Chironomus tentans*	Hoke et al., 1993
Microtox®, *Ceriodaphnia dubia, Chironomus tentans, P. promelas*	Hoke et al., 1990
Colpidium, D. magna, Hyalella azteca, and *Pontoporeyia*	Munawar et al., 1989c
D. magna, H. azteca, Chironomus tentans, Gammarus lacustris, Hexagenia limbata	Nebeker et al., 1984
Ceriodaphnia dubia, Hyalella azteca, Diporeia sp., *Chironomus riparius, Hydrilla verticillata*	Nelson et al., 1993
Microtox®, *S. capricornutum, Panagrellus redivivus*	Ross and Henebry, 1989
Microtox®, AlgalFB, *Daphnia magna*	Santiago et al., 1993
Lumbriculus variegatus, Pimephales promelas, Hyalella azteca	Schubauer-Berigan and Ankley, 1991; Ankley et al., 1991a
D. magna, H. azteca, C. tentans	Suedel et al., 1993
D. magna, Ceriodaphnia dubia, H. azteca, Chironomus tentans, P. promelas	Suedel et al., 1993
H. azteca, C. riparius, L. variegatus	Ingersoll et al., 1995
C. riparius, H. azteca, Hexagenia limbata, Tubifex tubifex	Reynoldson et al., 1995

contaminants in aquatic ecosystems at the population and community level in an attempt to reflect field conditions. Multispecies toxicity tests do not replace single-species bioassays but deal with a biological organization level higher than organism. Microcosms are either laboratory or outdoor bioassays conducted with several liters of natural sediment in small aquaria or other artificial vessels where several species, usually of different trophic levels, are exposed using a population density similar to that in the field. Toxicity responses in microcosm bioassays can be different than those observed when the species are exposed individually. Thus, Keilty et al. (1988a) found that toxicity to *Limnodrilus hoffmeisteri* was reduced by the presence of *Stylodrilus heringianus* in a mixed test system. Similarly, *L. hoffmeisteri* showed reduced postexperimental mortalities and increased biomass when mixed with *S. heringianus* in sediments dosed with endrin (Keilty et al., 1988b). Several microcosm tests have been proposed with different degrees of standardization, most comprising a representative species from several trophic levels, such as bacteria, protozoa, algae, and invertebrates. Cairns (1985) addressed several related problems with the multispecies testing and application; more recently, other contributions reviewed the multispecies testing experimental design, end points, and need for future research (Buikema and Voshell, 1993; Cairns and Cherry, 1993; Kennedy et al., 1995). Calow (1989) has identified the following problems related to multispecies testing for bioassessment: unstable artificial systems, their uniqueness makes the comparison of responses of different multispecies systems to the same toxicants difficult, and the cost is several times greater than the single-species assays.

Mesocosms are considered laboratory bioassays conducted at large scale. However, they are usually carried out in ponds or stream channels that more closely reflect conditions in the field. Because of their size, replicates and control series are limited. Mesocosm systems allow the measurement of secondary ecological effects such as predation, grazing, energy flow, food and habitat availability, recovery from perturbations, and long-term effects at the community level, as well as the fate of toxic substances (Buikema and Voshell, 1993). However, more research is still required to understand colonization and succession in mesocosms, for valid sampling and replication of measures of both structural and functional variables, which reflect cause–effect relationships.

There is no agreement on the replicability and standardization needs of multispecies bioassays. Some authors stress the use of the same species, exposed in the same number and physiological state, with their associated microbial biota, in a chemically defined medium and sediment (for example, Standardized Aquatic Microcosm) (Taub, 1989). Other sediments do not use the same community of organism, because they are more concerned about the ability to model particular interactions or responses at the population level than about standardization. Multispecies tests have been adopted by the U.S. Environmental Protection Agency (U.S. EPA) for ecological risk evaluation of pesticides in a tiered bioassessment approach prior to field testing (Touart, 1988). However, the tests still show considerable limitations in use for regulatory purposes.

For routine sediment bioassessments, the authors would recommend that a battery of single-species tests is most appropriate and cost effective. While emphasis should be placed on sediment infaunal test species, some other tests have been shown to be well correlated with them. The use of microcosms and mesocosms can be useful in some circumstances, as a high-level tier in the bioassessment approach, but are of limited value in routine assessment studies.

3.2.2 Sediment phase

Different sediment phases can be used in laboratory bioassessment. The solid phase can be whole (bulk) sediment, a whole sediment from the investigated site diluted with a clean sediment from an uncontaminated site, or an artificial sediment. The aqueous phase of the sediment may be extractable (soluble other than in water), water elutriate (i.e., water-extractable fraction), or sediment pore water. A decision on which sediment phase (aqueous versus solid) will be used for testing depends on the study objectives and will affect the selection of the indicator species.

The earlier sediment assessment literature describes mostly aqueous-phase testing. This was primarily because of the lack of standard solid-phase bioassays and the availability of testing protocols developed for effluents which could be readily used in sediment bioassessment. Testing of sediment elutriates was developed for the laboratory assessment of risk associated with dredging activity (Burton, 1991). These tests best reproduce the behavior of toxicants when sediments are resuspended in the water column during dredging and dredged material disposal. The tests can also be used for bioassessment of effects of any infrastructure works in rivers, lakes, and reservoirs (i.e., channelization works, dams) resulting in resuspension of sediments. Several techniques have been described to reproduce this phenomenon in the laboratory (Chilton, 1991; Schmidt-Dallmer et al., 1992). The toxicity of the suspended particles of sediment in the water column can also be measured to estimate the flux of pollutants and short-term changes in water quality. The sediment particles can be separated by centrifugation of large volumes of water (Santiago et al., 1993).

When rapid screening is required, bioassays on pore water may be adequate (Hill et al., 1993). However, there are no standardized extraction methods for this sediment phase, and the methods vary from author to author, making comparisons of results difficult. Furthermore, evaluation of the toxicity of sediment pore water is difficult, as there is little knowledge on the effects of its chemical and physical properties on the organisms, in addition to the contaminants present. For example, Hoke et al. (1992a) demonstrated that high alkalinity of pore waters from polluted sediments of Grand Calumet River produced acute toxicity to *Daphnia magna* and *Ceriodaphnia dubia*, suggesting that it may be a potential confounding factor for invertebrate tests.

The general trend in sediment assessment is to use bulk sediment, and new protocols have been proposed in the last few years to facilitate this approach (e.g., Borgmann and Munawar, 1989; Brouwer et al., 1990; Reynoldson et al., 1991). Bulk sediment bioassays best represent the conditions of exposure of organisms in the field. However, as most of the protocols being proposed are quite recent, more research is still needed, particularly on intercalibration and validation of the methods. Despite this fact, 54% (i.e., 75% of invertebrate testing) of the reviewed literature on sediment bioassessment published in the last 8 years uses solid-phase bioassays, although most still combine both aqueous extracts and whole-sediment bioassays in a battery of tests.

There are few studies which compare the results of toxicity tests for organisms exposed to different phases (Munawar et al., 1989c; Ankley et al., 1991b; Harkey et al., 1994). In most cases, sediment pore water has been shown to be more toxic than elutriates from the same sediment and in some cases more toxic than the bulk sediment. Toxicity data generated using pore water have also been shown to correlate better to bulk sediment toxicity data than data generated using sediment elutriates (Schubauer-Berigan and Ankley, 1991).

The use of bulk sediment bioassays involves less sediment handling, and it is, in the opinion of the authors, more reliable than liquid-phase testing. A sufficient number of bioassays have been developed using the solid phase to be included in a battery of bioassays on organisms from bacteria to fish. However, more attention should be paid to intercalibration and good validity criteria of these bioassays.

3.2.3 *Exposure time: acute versus chronic bioassays*

Most sediment bioassessments have been conducted using acute toxicity bioassays, with short-term (i.e., up to 96 hr) exposure periods ranging from 15 min in the Microtox® bacteria test to 48 hr for cladocera tests and 96 hr for macroinvertebrate and fish tests. These tests have poor discrimination power for classification purposes, but do produce rapid responses for legal decisions when high levels of pollution exist. However, because most sediments or extracts do not provoke a lethal response in such a short-term exposure, long-term (i.e., more than 96 hr) chronic bioassays measuring survival and impairment of growth and/or reproduction in long-term exposures have become common in most sediment bioassays. The chronic bioassays require days to months of exposure (i.e., usually more than 4 days) and include three types of sublethal responses (Buikema and Voshell, 1993): (1)

life cycle (i.e., a generation period, from egg to egg), (2) life history (i.e., from egg to death or cessation of reproduction), and (3) sublife cycle (i.e., a period comprising a relatively short sensitive life stage, commonly of newly hatched animals), usually referred to as subchronic tests. Many batteries of bioassays combine both acute and chronic measures in different species. Efforts have also been made in recent years to develop subchronic tests that, if highly correlated with the chronic responses, can provide more rapid and less expensive information. Chronic and subchronic standardized sediment bioassays are highly recommended for classification of sediments in areas of concern, for remediation programs, and to establish quality criteria for aquatic sediments.

3.2.4 *Selection of relevant end points*

In bioassays, toxicity is defined by a preselected adverse effect termed an end point (Luoma and Carter, 1993). The selection of a relevant end point should be consistent with the objectives of the survey and the degree of contamination of the sediment in the study area. Thus, a first screening of very polluted areas can be undertaken with simple, rapid, and economic bioassays using survival or immobilization of invertebrates, bacterial luminescence, or growth inhibition bioassay. However, for classifying large areas, conducting long-term remediation programs, and establishing biological sediment objectives, chronic or subchronic physiological, ecologically relevant end points are preferred. While qualitative responses (i.e., yes or no), exclusively based on survival measurements, are useful for rapidly required legal decisions, they can be misleading and do not allow classification of sediments into ranked categories from clean to polluted. Some recent work focused on functional continuous response variables, such as growth or reproduction inhibition, which increase the discriminatory power of the bioassay (Giesy and Hoke, 1989). Growth of the test organisms is a good integrative measurement of the individual response to contaminants (Power et al., 1991) and shows a dose–response dependency that has been widely referenced. Somatic growth alteration, relative to control conditions, has consequences for reproduction rates. A reduced growth rate implies reduced reproductive output in terms of biomass (not always in the number of eggs) that can result in the inability to form gonads or the nonviability of embryos and therefore difficulty of populations to sustain themselves.

In a review of chronic tests by Power et al. (1991), reproduction has been found to be a more sensitive response than either survival or growth. Most bioassays measure the number of eggs or juveniles produced during a long-term exposure as reproductive end points. However, effects on reproduction may include delay in sexual maturation, brood release, egg development, hatching, and biomass of the brood, as well as total inhibition of sexual reproduction. The number of juveniles produced is a complex end point that results from a combination of reproductive output and young survival, thus reflecting an effect at a population level more than reproductive impairment at an individual level.

A desirable attribute of any end point is that it should have a low coefficient of variation to enhance its reliability in bioassessment. Another consideration in end point selection is specificity of response. Those that respond very generally to all toxicants are useful for screening when a causative agent is unknown, while end points that respond specifically to a narrow range of contaminants are useful for diagnosis (Cairns et al., 1992).

Behavioral (such as avoidance responses or burrowing activity), cellular, and biochemical end points are rapid, sensitive at low contaminant levels, and operate at a lower level of organization in the organism. However, results must be supported by other end points of ecological relevance (i.e., affecting individual or population levels) for management action to be taken (e.g., Coler et al., 1988). Keilty et al. (1988a) compared lethal and behavioral responses (sediment avoidance) and found behavioral responses at values 46 to 150 times lower. *Gammarus pulex* "drift" has also been described as dose dependent, and behavioral changes are identified at concentrations where lethal responses were achieved in prolonged exposures (Taylor et al., 1994). However, the definition of behavioral end points is generally imprecise, rendering intercomparisons difficult, but they can be useful as a warning of stress at low concentrations of toxicants. Although very sensitive, cellular and biochemical changes may be of short duration or reversible without exerting impact at the community level (Power et al., 1991). Changes at this level may simply be a response to homeostatic adjustments in the organism and may have no effect on the survival, development, and fecundity of the individual and thus be of doubtful ecological relevance (Calow, 1989). For sediment bioassessment, the choice of end points that measure primary effects of concern (i.e., lethality, reproduction, or growth inhibition) is recommended, avoiding substitution by rapid bioassays (biochemical or cellular) that should be used as early warning systems but not for diagnostic information (Chapman, 1995a).

Multispecies tests utilize both physico-chemical parameters (i.e., sediment oxygen demand, pH, Eh) and biological end points (i.e., predation–prey interactions, respiration, production, community structure) and have been reviewed by Cairns and Cherry (1993). Examples of common structural (species richness and abundance and biomass) and functional (community production and respiration, rates of decomposition, nutrient uptake/regeneration) end points are provided by Cairns et al. (1992).

Selection of suitable end points, rejecting those that provide redundant information or are highly variable, or including some rapid cellular or biochemical end points, as well as establishing correlations with other more relevant bioassays, may require an extensive literature review or can be addressed in a pilot study.

3.2.5 Selection of the test organism

Criteria for the selection of test organisms have been discussed by several authors (Giesy and Hoke, 1989; Burton et al., 1992; Hill et al., 1993. Table 3-2 summarizes those proposed in the literature. The initial decision to be made when selecting test organisms is whether to use surrogate or sediment-associated species. Surrogate species are sensitive standardized species, such as cladocerans or the *Photobacterium* bacteria, that show methodological advantages and are readily cultured under laboratory conditions. They have well-established protocols, and there is considerable information in the literature on their response to contaminants. In addition, interlaboratory comparisons are relatively straightforward. Surrogate species have mostly been used for evaluating acute end points on survival or other sublethal effects. They are useful for characterizing relative effects in a specific area or a discharge plume. However, data from a standardized species cannot predict the responses of indigenous species (Chapman, 1991). The paradigm "that protection of the most sensitive species will inadvertently protect all other" has been criticized as unrealistic: different species have different sensitivities to different chemicals (Cairns and Pratt, 1993).

Sediment-associated species allow better prediction of field sediment quality bioassessment because they have more ecological relevance, which facilitates extrapolation from laboratory to field conditions (Chapman, 1995a). In general, organisms cultured in the laboratory are preferred over those taken from the field as they allow a greater degree of standardization (Hill et al., 1993). Therefore, the choice of species is, in most cases, not related to their "importance" in the ecosystem but to pragmatic factors such as the availability of a sufficient number of individuals for bioassay, tolerance to laboratory conditions, and ease of culture and handling. However, these species can be calibrated against known sensitive species to determine their relative sensitivity (Chapman, 1995a).

Table 3-2 Criteria for the Selection of Test Species

General biological criteria
Sensitivity relative to other species and to different chemicals
Ecological relevance
Exposure routes to toxicants relative to the habitat (infaunal, epibenthic, planktonic, etc.)
Database on the biology of the species

Laboratory criteria
Toxicity testing protocols
Ease of culture and handling, low mortality in laboratory conditions
Suitable end points: relevant, consistent (not highly variable), cost effective
Tolerance to a range of different sediment matrices (particle size, organic content)

For interlaboratory comparisons, organisms from the same culture stock or the same locality should be used or, alternatively, different populations should be intercalibrated using reference toxicants or by conducting bioassays in parallel for comparison of relative sensitivities to measure several chronic variables (e.g., Reynoldson et al., 1996). If bioassay organisms have to be collected from the field, information on the sediment chemistry at the site has to be provided, as some populations are able to adapt to low-level long-term stresses (i.e., chemical or otherwise) and show lower sensitivity to pollutants (Keilty and Landrum, 1990).

Finally, benthic infaunal species should not be used when conducting aqueous-phase bioassays (Burton et al., 1989; Harkey et al., 1994), as the absence of sediment may be an additional stress, resulting in an overestimation of the response to toxicants, although this consideration will be of concern mainly when chronic bioassays are to be conducted. We favor the use of species naturally living in sediment for the bioassessment. However, a battery of species with both these and surrogate species can also be used depending on laboratory facilities.

3.3 *Sampling decisions*

The ASTM (1990), the Canadian Council of Ministers of the Environment (1993), and Hill et al. (1993) provide good advice on sampling issues, including number of sites, replicates, which sampling device is most adequate, and issues related to handling and storage of sediments. There is no simple formula for designing a sediment sampling program which would be applicable to all sediment bioassessment studies. Therefore, when an expensive/extensive survey is to be carried out, many strategic sampling issues should be addressed in a quick pilot study on a few sites. This information will assist in making a decision on many sampling issues and will avoid the production of data that are either methodologically inadequate, statistically unacceptable, or of either too fine or coarse resolution for the study objectives. Most bioassessment surveys stand on a unique sampling episode; however, there has been some demonstration that sediment toxicity shows a temporal variance (Burton, 1989; Munawar et al., 1989c; Borgmann and Norwood, 1993). While this may not have important implications if the bioassays are to be used for classification purposes, by comparing sediments from several localities sampled in the same period, it may be important when evaluating remediation programs or hazard from sediment containing contaminants from effluent discharges and may require several sample events through the year(s).

3.3.1 *Selection of sampling locations*

Decisions on the site selection again depend on the objectives of the study, the ecosystem being investigated, and the available budget. MacKnight

(1994) emphasizes that there is no single formula for the design of a sediment sampling pattern that is applicable in all sediment sampling programs. A haphazard pattern is commonly used, where sites are located based on ease of access, ease of sampling, and proximity to the areas of concern, such as industries or towns. This design may be inadequate for screening and remediation programs where a grid or block sampling design allows an unbiased and random selection of sites and facilitates statistical treatment of results identifying trends in the effects of toxicants. Bioassessment studies that endeavor to measure the effect of point-source effluents or any other source of toxicants can locate sampling sites on a transect at geometrically growing distances. River studies should concentrate on sampling of depositional areas of sediments, ease of access into a grid, and avoiding rapids and riffles where erosion can transport fine particles downstream quickly.

3.3.2 *Sampling devices*

Mudroch and Azcue (1995) describe a wide variety of sediment samplers. Elliott and Tullett (1978) provide a comprehensive bibliography of sampling devices. Choosing the most appropriate sediment sampling device will depend on the sediment characteristics, the volume and efficiency required, and the objective of the study. The strengths and weaknesses of different sampling devices for sediment bioassessment are discussed by the ASTM (1990) and Burton (1992a). Ekman and Ponar grabs and different corers are the most commonly used sampling devices. They are recommended by the ASTM (1994). Models made of stainless steel are recommended in particular for pollution studies, although problems related to metal contamination from samplers can be avoided by using plastic liners in corers.

A comparison of the efficiency of core and Ponar dredge samplers for identifying historical contamination and bioassessment in a stream system polluted with DDT showed that cores provided more information. However, sediment toxicity to *Chironomus tentans* in a 10-day exposure in sediment bioassays was similar regardless of the sampling method. Corers produce the least disturbance of the sediment and are more efficient than grabs (Bufflap and Allen, 1995). However, corers are not suitable for sandy sediments, and they remove only small quantities of sediment, usually requiring several core samples to be composited. The investigator should be aware of the possible difficulties that such manipulations of the sediment can have on meeting the survey objectives. For studies in sediment remediation by dredging and historical pollution studies, long corers up to several meters can be used. If available, box corers that remove portions of the bottom (up to 50 cm × 50 cm to a depth of 40 cm) are preferred as they remove an intact section of bottom material. However, this equipment requires a relatively large ship equipped with a crane, which is not usually available. A smaller version of this device, developed at the Canada Centre for Inland Waters, removes a section of bottom sediment 25 cm × 25 cm and up to 20 cm depth and can be used on a relatively small boat with a winch.

3.3.3 *Number of sites, sample volume, and replicates*

The number of sampling sites depends on the study objectives and also on the distribution and nature of pollutants, laboratory capabilities, and costs. However, some work was carried out to establish a minimum necessary number. Håkanson (1992) demonstrated that the number of sampling sites is a function of the area, the shore development, and variability of organic matter content in the surficial sediment (i.e., loss on ignition).

There are two major forms of replication: replicating field samples and laboratory replication. Field replication requires discrete samples to be taken at a site in the field and tested separately with or without laboratory replication. Laboratory replication involves the splitting of a single sample in the laboratory and performing replicate tests. If only field replication is carried out, then the resulting data include variation due to spatial variability and test variability. If only laboratory replication is carried out, then the variability is only due to inherent test variability (i.e., organism and laboratory procedures). It is advisable when undertaking any new test methodology to include both types of replication to establish which source of error is greater. Several authors have replicated field samples to measure the spatial heterogeneity of the sediment. The number of replicates required depends on the degree of horizontal variance, which is a site-specific function. It varies with the nature of the sediment and its interaction with the toxicants present, and it is advisable to establish the number of replicates required when assessing sediment quality (Stemmer et al., 1990b). A pilot survey can establish the appropriate number of field replicates for a required degree of variation or precision of the data.

Depending on the spatial scale of the survey and the spatial resolution required, compositing replicate samples to achieve adequate representation of the sampling location may be necessary. For example, if the remedial option for contaminated sediments is dredging, the spatial resolution is determined by the dredging equipment (for example, 20 m × 20 m = 400 m^2). Therefore, a sampling grid finer than this is unnecessary and compositing samples into a block may be an option. The upper 5 cm represents the most biologically active portion of the sediment. This portion can be removed with cores or scoops from a grab sampler. It is the appropriate depth for most assessment objectives, except historical and dredging issues that require information from deeper sediment layers. Another common approach is to take discrete multiple grab samples from a site and composite and homogenize them in the field. However, if the subsurface sediment layer has a lower redox potential than the top of the sediment, this practice should be avoided, because it can result in an underestimation of sediment toxicity, as oxidized metal complexes are less available (Burton and MacPherson, 1995). Most important is to use standardized or well-evaluated methods throughout the study.

3.3.4 *Reference sites and control sediments*

Knowledge of the responses of different benthic species to natural sedimentary environments is required to avoid "false-positive" results when interpreting test responses. Thus, the investigator must be able to discriminate adverse effects due to pollutants from stress due to natural conditions. Reference sediments have been defined as sediments that possess characteristics similar to the test sediment, such as particle size distribution and concentration of organic carbon, but without anthropogenic contaminants (Burton, 1992a). Results from unpolluted reference sites of various types can provide an estimate of the normal response that can be expected from an organism exposed to sediments. Reynoldson et al. (1995) showed that there is variation of some test end points depending on natural sediment characteristics, the species of invertebrate used in the bioassay, and the response being measured. For example, the response of *C. riparius* to sediment type is very robust and neither growth nor survival is notably affected by sediment quality or nutrition. *Hyalella azteca* and *Hexagenia* sp. growth measurements were quite variable.

Hughes (1994) discussed criteria for selecting reference sites and the acceptability of sites for ecosystem classification that can be applied to a sediment bioassessment program. In rivers and streams, rapid screening of the benthic invertebrate families present at a site and the measurement of basic chemical variables can give a good approximation of an unpolluted reference state. However, a complete chemical analysis of the sediments to discount pollution and similar particle size distribution of organic matter content of the test sediments are desirable.

To provide evidence on the health of the test organisms, for determining adequate handling and test conditions, for the application of validity criteria, and for statistical comparison of the level of response in organisms exposed to polluted versus unpolluted sediments, it is necessary to establish a control sediment series as part of the bioassays. This is usually carried out with the sediment used for culturing the test organisms or a reference sediment with known responses for the end points to be measured. Rarely have artificial sediments been used only as a control for bulk sediment bioassays, for example, autoclaved sand described by Schubauer-Berigan and Ankley (1991). In any case, it is preferable that the control sediment(s) have characteristics similar to the test sediments, such as the concentration of total organic carbon distribution, pH, etc. (ASTM, 1994).

3.3.5 *Sampling issues: storage time, temperature, materials, handling*

Factors during sediment sample collection and processing that can affect integrity and characteristics of sediments have been reviewed and discussed by Burton (1992a). Sample containers commonly used are made of glass and high-density polyethylene, to avoid leaching, dissolution, or sorption of chemicals. Different apparatus for manipulation of the sample can be made of stainless steel, plastic (i.e., polycarbonate, polyethylene, and fluorocarbon), or Teflon. It is interesting to note that contamination of water samples is by far a more serious problem than contamination of sediment samples by the material of containers (Mudroch and Bourbonniere, 1994).

Samples are commonly transported to the laboratory on ice and kept in the dark at 4°C until used for bioassays. It is generally recommended that samples be tested as soon as possible, but this is not always practicable. A conservative storage time of 2 weeks for whole sediment is recommended (ASTM, 1994; Burton and MacPherson, 1995). However, several authors have shown longer storage times, such as 3 to 16 weeks (Ankley et al., 1990, 1992a) and less than 5 weeks (Borgmann and Norwood, 1993). Some authors reported no statistical differences in acute or chronic toxicity after the storage period, for example, 6 months (Reynoldson et al., 1991) and 112 days (Othoudt et al., 1991). Becker and Ginn (1995) analyzed the effect of a storage time of 16 weeks in three different sediment toxicity bioassays and showed that differences in toxicity levels after the storage period were greater for those sediments that showed low to intermediate levels of toxicity than for those indicative of high toxicity. For water extracts, storage times are very critical, and many authors considered it necessary to test the aqueous phases immediately or in less than 24 hr after extraction, as water chemistry can change significantly (Burton, 1991). Freezing is not recommended as it reduces the toxicity of the overlying water (Malueg et al., 1986). The use of a glove box with an inert atmosphere may be necessary for processing sediments (i.e., subsampling and processing) to avoid oxidation of anoxic layers (Burton, 1992a) for some specific tests, but it is rarely used for routine bioassays.

3.4 *Preparation of the bioassay*

3.4.1 *Aqueous-phase bioassay*

3.4.1.1 *Pore water extraction*

Adams (1991) and Bufflap and Allen (1995) recently revised pore water extraction methods, some of which are used in toxicity bioassessment. The ASTM (1994) considers six methods of extracting pore water for conducting

Table 3-3 Methods Reported for Pore Water Extraction in Sediment Toxicity Bioassessment

Pore water extraction method	Storage time	Reference
Centrifuge 4000 g (45 min), refrigerated + 24-hr settling + decantation	<72 hr	Ankley et al., 1990
Centrifuge 2500 g (30 min), 4°C	—	Ankley et al., 1991b
Centrifuge 8000 rpm (45 min), 20°C + filtration 1.2 μm	< 7 days	Giesy and Hoke, 1989; Giesy et al., 1988a
Centrifuge 4000 g (30 min), 10°C + supernatant 2000 g (30 min)	<24 hr	Harkey et al., 1994
Centrifuge 7000 rpm (60 min) + filtration 1.2 μm	<7 days	Hoke et al., 1990
Centrifuge 6730 g (1 hr), 4°C + decantation	<7 days	Kubitz et al., 1995
Centrifuge 9000 rpm (15 min), 4°C	immediately used	Sasson-Brickson and Burton, 1991
Centrifuge 2500 g (30 min), 4°C	<72 hr	Schubauer-Berigan and Ankley, 1991

sediment toxicity tests, but centrifugation is the most widely used methodology, as it is quick (usually 30 min can be sufficient) and provides a sufficient volume of interstitial water. Ankley et al. (1991b) have indicated that centrifugation is preferable to squeezing or syringe extraction for pore water isolation. Another advantage is that centrifugation can be performed under refrigerated conditions to prevent artifacts due to temperature changes. The appropriate centrifugation speed and time have to be established for each sediment and are quite variable in the literature (Table 3-3). Subsequent filtration may be necessary if elimination of suspended particles is required. However, this is an extra handling step that is preferably avoided (Schubauer-Berigan and Ankley, 1991), as it has been demonstrated to remove contaminants and significantly reduce toxicity (Sasson-Brickson and Burton, 1991). Centrifugation at high speeds (7600 g) or settling followed by decantation is more adequate for the elimination of suspended particles in the supernatant water (Burton, 1991).

3.4.1.2 Elutriate preparation

The technique of elutriation provides a more standardized approach for the extraction of water-soluble chemicals. Elutriation is achieved by mixing sediment and water (hard or very hard reconstituted water, site water, or Milli Q) at a ratio of four parts water to one part sediment (by volume),

Table 3-4 Methods Reported for Elutriation (Extraction of Water-Soluble Fraction) in Sediment Toxicity Bioassessment

Elutriation techniques	Reference
1:4 s/w[a] + mixed 1 hr + centrifugate 2500 g (30 min), 4°C	Ankley et al., 1991b
1:4 s/w + shaken overnight + homogenized in a blender at 21,500 rpm (5 min) + centrifuge 5000 g (5 min)	Bitton et al., 1992
1:4 s/w + mixed 30 min + centrifugate + filtrate 0.45 μm for algal growth tests	Burton, 1991
1:1 s/w + mixed 3 min + refrigerated centrifuge 5000 rpm (10 min)	Dutka and Kwan, 1988
1:1 s/w + mixed 2 min + centrifugate 10,000 rpm (20 min)	Dutka et al., 1989
1:4 s/w + mixed 30 min + settling 1 hr + centrifuge 2000 g (30 min), 10°C	Harkey et al., 1994
1:4 s/w + mixed 1 hr + centrifuge 10,000 rpm (30 min) + filtered 0.45 μm	Munawar et al., 1989c
1:4 s/w + mixed 10 min + centrifugate 10,000 rpm (15 min)	Nebeker et al., 1984
1:4 s/w + mixed + centrifuge 15,000 rpm + filtered 0.45 μm	Nebeker et al., 1988
1:4 s/w + mixed 1 hr + supernatant filtered 0.45 μm	Santiago et al., 1993
1:4 s/w + mixed 30 min + centrifuge 9000 rpm (15 min), 4°C	Sasson-Brickson and Burton, 1991
1:4 s/w + mixed 1 hr + centrifuge 2500 g (30 min), 4°C	Schubauer-Berigan and Ankley, 1991

[a] Sediment/water ratio.

followed by stirring, shaking, or rolling for 30 min to 24 hr. The water phase is then extracted after settling of the particles either by decantation, centrifugation, and/or filtration (Table 3-4). If extracts of the organic soluble phase are required, they can be obtained by adding solvents, such as dichloromethylene, to the sediment (Dutka and Kwan, 1988; Santiago et al., 1993). This is an approach to assess the toxic agents in complex mixtures. The efficiency of several sediment/water ratios and mixing times was tested by Daniels et al. (1989) through physico-chemical analysis and ^{14}C uptake by *Chlorella vulgaris*. The authors showed that the most efficient extraction was a mixture of 1:4 sediment/water and a mixing period of 1 hr. These authors also compared different mixing systems and described an improved elutriation technique using rotary tumbling for sediment/water mixing that exerts no excessive mechanical stress or major changes due to the oxidation of the sediments. Where only small quantities of sediments are available for elutriation, Munawar et al. (1989a) have designed a procedure called Limited Sediment Bioassay.

3.4.1.3 Dose–response water extract tests

There are numerous references to bioassays conducted on aquatic phases for effluents and receiving waters (e.g., U.S. EPA, 1989, 1991) that can be applied to sediment/water extracts. Typically, dose–response bioassays generate information on the proportion of individuals being affected by a known level of dilution of the water extract (for example, EC_{50}) or on the No-Effect Level that can be used for screening and classification of sediments from unpolluted to highly polluted or for sensitivity comparison of the test species.

3.4.1.4 Conclusions on water-phase testing

Aqueous sediment extracts do not accurately represent the exposure observed in the test with bulk sediments (Harkey et al., 1994). The fundamental problem with all aqueous-phase testing is the lack of reality. The availability of contaminants in sediments is largely determined by the sediment matrix in which the biota reside, and any extraction method inevitably destroys the relationship between the contaminant and the sediment, providing unrealistic results. Therefore, except for very specific cases, such as species sensitivity comparisons and dredging studies, the authors would recommend the use of bulk sediment bioassays.

3.4.2 Bulk sediment bioassays

Detailed descriptions of bioassay methods on bulk sediments are provided by the ASTM (1994) and other authors. Therefore, no attempt is made to describe each existing test here. However, some general discussion on the preparation required to test a whole sediment and some discussion of current methodological issues are provided.

A sufficient quantity of sediment from the field replicates, or composited samples, should be homogenized to conduct the required number of laboratory replicates. The basic experimental design for sediment bioassessment consists of a bulk sediment test series of replicates run with a control sediment series. The necessary temperature, dissolved oxygen levels, and settling of the suspended solids should be achieved before the introduction of the test organisms to the test. This is usually carried out by allowing the test chambers to stand for a period, usually 24 hr, under test conditions before the introduction of the test organisms. Oxygen should be measured prior to adding the test organisms, and if lower than a predetermined level, which depends on the test species, the water overlying the sediment in the test chambers should be aerated. The overlying water that evaporates during the bioassay should be replaced by deionized or distilled water. If oxygen levels go below the tolerance limits of the species, aeration in static systems has to be provided. The oxygen should be introduced at a slow rate to avoid resuspension of the surficial sediment layer.

3.4.2.1 Dose–response tests

Polluted sediments can be diluted to produce a concentration gradient of contamination for estimating the EC_{50} and a no-effect dilution level. While useful for setting and verifying concentrations of contaminants for cleanup of the sediments, there are problems with this approach. One should be aware that all dilution techniques disrupt sediment integrity and probably affect the availability of chemicals in the sediments. The first decision to be made is the choice of the diluent. Water has rarely been used (e.g., natural water [Wiederholm and Dave, 1989], reconstituted water [Dave, 1992a,b], pore water [Giesy et al., 1990]), whereas sediment (e.g., natural or artificial such as sand or a silt–clay soil) is more common. If natural sediment is used as a diluent, we would recommend using a clean sediment with geochemical characteristics similar to the test sediment, particularly with respect to organic matter content and particle size distribution (Burton, 1992a; Hill et al., 1993). The effect of sediment as a diluent and any consequent biological amelioration is more than simple dilution because of the increased fresh sites on sediment particles for sorption of contaminants (Burton, 1991). This approach is more realistic and better understood when considering the effect of sediment dilution as part of the natural processes by the settling of clean sediment particles on the surface of polluted sediments. Nelson et al. (1993) compared several sediment diluents and found that a heterogeneous natural sediment reduced toxicity of the test sediment more effectively than sand, as sand provides minimal contaminant sorptive properties relative to fine silt–clay particles. When mixing the sediments, appropriate methods and periods of equilibration must be used. A rolling mill is generally appropriate for homogenizing the mixed sediments before dispensing aliquots for the bioassay replicates, although mixing by hand with clean tools can be sufficient. An equilibration period is necessary, but it is quite variable, extending from several hours (Nelson et al., 1993) to at least 10 days (Giesy et al., 1990). Sediments can be mixed on a volume-to-volume or wet weight basis, although the former appears to be superior (Burton, 1991).

3.4.2.2 Spiked sediments

The addition of individual contaminants to sediment can also be used for measuring a dose–response relationship of chemicals associated with sediment. Spiked sediment bioassays allow estimation of the EC_{50} or the No-Effect Levels for single substances or mixtures, as well as bioaccumulation factors, for deriving sediment quality standards and for predicting hazards from effluents or polluted sediments.

Hill et al. (1993) and the ASTM (1994) provided a detailed description of spiking procedures for both water-soluble and insoluble substances. The three principal methods for spiking are (1) addition of the chemical to the overlying water (Burton et al., 1987), (2) addition of a solution of the chemicals to the wet sediment (Ditsworth et al., 1990; Landrum and Faust, 1994; Landrum et al., 1994), and (3) "shell coating" the walls of the jar with the chosen concentration of the toxicant (for nonpolar organics) by addition of the solution to the test jar and allowing the carrier solvent to evaporate. After this, the sediment is added and rolled for several days (Ditsworth et al., 1990; Hoke et al., 1995). Other methods of addition can be considered, depending on the objectives of the bioassays, for instance, spraying of pesticides as used in forestry or agricultural practices.

There is no standardization of mixing systems, mixing periods, and time for equilibration, all of which are important factors that affect the equilibrium partitioning and consequently the bioavailability of the toxicants (Burton, 1991). Current guidance recommends minimal mixing times on the order of minutes to a few hours to reduce oxidation of sediments (Stephenson et al., 1995). Due to the lack of standardization, the homogeneity of the mixed sediments should be demonstrated. There is little information about the volume of carriers used in the spiking of organic chemicals, which may be contaminants in themselves. Therefore, the carriers have to be added at minimum volume and also to the bioassay control series. Whatever spiking method is used, once a concentration series has been prepared, it is essential to measure the real concentration of the chemical in the stock solution, in bulk sediment, and in overlying waters after the spiking procedure is finished, before adding the organisms (ASTM, 1994). This is very important because initial concentrations of the spiked chemical can be quite different from the final concentrations in the spiked sediment, depending on the method used. Furthermore, the concentration of spiked chemical can also be different in pore water, and sediment/water partitioning of the chemical needs to be established (Burton and MacPherson, 1995). Stemmer et al. (1990a) investigated several methodological issues related to spiking procedures. They found similar toxicity effects of selenite-spiked sediments to *Daphnia magna* using different mixers. They also proved that mixing time dramatically reduced mortality after 24 hr and that spiked sediments kept at 4°C showed similar toxicity effects for 2 days. Finally, a decrease from 1:4 to 1:8 sediment/water mixture as well as an increased surface area of the sediment decreased the survival of test organisms.

Once the sediment has been spiked with chemicals, an equilibration time of several days to weeks is required, depending on the spiked chemical and the nature of the sediment. An approximation to this time is given by measuring the concentration of the spiked chemical in pore water until no changes are registered (Stephenson et al., 1995). The exposure of test organisms can be conducted using spiked bulk sediment, elutriates, or pore water extracts (Harkey et al., 1994; Kubitz et al., 1995).

Table 3-5 Components of Artificial Sediment for Aquatic Toxicity Testing

Components of the artificial sediment	Reference
10–20% fine (<63 μm) + median size (90–125 μm) + 0.5–4% organic matter	Hill et al., 1993
70% fine sand (50% 50–200 μm) + 20% kaolinite clay + 10% sphagnum peat + CO_3Ca to pH = 6.0 ± 0.5	Organization for Economic Cooperation and Development, 1983, 207 guideline
76% quartz sand + 22% kaolinite clay + 2% sphagnum peat + 46% water	Egeler et al., 1995
8% coarse sand + 32% medium sand + 36% fine sand + 21% silt + 3% clay + humus for 0.1% organic matter	Suedel and Rodgers, 1994
75% clay + 25% sand; 25% clay + 75% sand; 50% clay + 50% sand; 33% clay + 33% sand + 33% silt	Stephenson et al., 1995

3.4.2.3 *Artificial or formulated sediments*

Research on artificial sediments is important as they may allow more standardization between laboratories for spiking chemicals and dilution of sediments. This approach will also focus on the role of specific sediment components, such as the particle size distribution, in the toxicity responses of organisms. While it is not possible to describe an artificial sediment that is optimal for every test species, as this requires knowledge of the tolerance of species to a particle size range and, in particular, to the presence of organic matter, it is possible to demonstrate that control conditions allow acceptable values for the test end points. Table 3-5 summarizes some recently proposed formulated sediments for bioassays.

3.4.3 *Methodological issues*

3.4.3.1 *Removal of indigenous fauna*

Indigenous invertebrates present in the natural sediment can interfere with a particular end point, such as survival through predation, growth through competition for food, or reproduction of organisms other than the test species. For some simple acute bioassays, removal of predators or large animals by hand can be done with minimal disruption of sediment. Interferences by indigenous oligochaete worms in the sediments have been reported by Dermott and Munawar (1992). Several bioassay protocols include sieving of sediments to eliminate indigenous organisms (Milbrink, 1987; Borgmann and Munawar, 1989; Rosiu et al., 1989; Reynoldson et al., 1991), with mesh size varying between 250 mm and 1 or 2 mm. Other authors strongly discourage sieving the sediment to avoid changes in the toxicity of sediments, due to oxidation and concomitant release of metals as well as changes in particle size distribution (Phipps et al., 1993). However, Reynoldson et al. (1994) showed the effects of indigenous organisms in sediment toxicity tests conducted with three species of benthic invertebrates: *Chironomus riparius*, *Hyalella azteca*, and *Hexagenia limbata*. Indigenous organisms were simulated by various densities of the oligochaete worm *Tubifex tubifex*. Worms did not affect survival of the test species but did reduce growth of all three test species and two species at the lowest densities tested, equivalent to 1460 individuals per square meter. At densities of *T. tubifex* equivalent to 20,000 per square meter, the growth of *C. riparius* was reduced by more than 90%, *Hyalella azteca* by more than 60%; and *Hexagenia limbata* by almost 50%. The densities of oligochaetes used were equivalent to those found in many contaminated sites. In a study of sediment manipulation methods for the removal of indigenous organisms, Day et al. (1995) examined the effects of sieving, freezing, autoclaving, and irradiating on both the chemistry of the sediment and results of three bioassays. Neither particle size distribution nor concentrations of metals, PCBs, or PAHs varied much due to manipulation. However, survival of *Hyalella azteca* and the growth of *C. riparius* were affected by autoclaving sediment, and young production in *T. tubifex* was enhanced by irradiation. Other methods of removing indigenous fauna from sediment, such as heating, drying, or radiating, were not recommended by Stephenson et al. (1995). Therefore, sieving through a fine mesh seems to be the method that least affects the toxicity response of organisms. This method is appropriate for bioassays if endemic organisms constitute a potential confounding factor.

3.4.3.2 Feeding

Acute (i.e., less than 4 days) bioassays do not require feeding the organisms. However, chronic and subchronic bioassays (i.e., more than 4 days) usually require feeding of the test animals to ensure acceptable control survival, growth rate, or reproduction. Yet, some authors exclude feeding, as it can mask toxicity effects and lead to underestimation of sediment toxicity. Instead, these authors suggest that the test sediment has to be renewed regularly (Milbrink, 1987). Several chronic bioassays require feeding every day during the test period or with every overlying water renewal in a semi-static test system, while others require feeding only at the beginning of the test. In static bioassays, the latter is preferred over feeding intermittently during the test, as it potentially produces less disturbance to the test system.

Reynoldson et al. (1995) showed that growth of *Hexagenia* spp. was highly variable in reference unpolluted sediments, and this variation was likely due to the fact that *Hexagenia* nymphs were not fed during the laboratory tests. The nutritive quality of the sediments is the limiting factor in the growth of the organisms. The same authors also suggested that a lack of exogenous food can explain the greater variability in the reproductive end points measured in the oligochaete test with *T. tubifex*. Total numbers of young oligochaetes have been shown to be sensitive to the amount of available food in the sediment, as measured by organic matter content (Reynoldson et al., 1991). Ankley et al. (1994a) studied the influence of feeding on invertebrate species. They found that the 10-day survival *Hyalella azteca* and 10-day survival and growth *C. tentans* bioassays had significant potential for false positives and failing control validity criteria if not fed. The nature of the supplementary food added to the test chambers depends on the organism. Algae or YCT (U.S. EPA, 1989) was used for feeding cladocera and as fish food for benthic invertebrates. However, the food should be in a minimal quantity that will not alter the organic content of the sediment.

3.4.3.3 Exposure conditions

The overlaying water can be maintained for the duration of the bioassay in a static system, or it can be partially renewed each day in a static-renewal system. To simulate field conditions, bioassays can be conducted in a continuous-flow regime. This is common for fish bioassays, but has also been used with invertebrates (Ankley et al., 1993a; Nelson et al., 1993; Hoke et al., 1995; Phipps et al., 1995). Benoit et al. (1993) recently described a portable intermittent renewal system for invertebrate toxicity bioassays that allows renewal of 1 to 21 volumes of water per day.

Some authors have suggested that static conditions overestimate toxicity and have advocated flow-through systems. However, Hill et al. (1993) recommended aeration of overlying water, if necessary, in a static system over flow-through or static-renewal systems. Sasson-Brickson and Burton (1991) compared *in situ* exposures with a laboratory static and flow-through water renewal system and showed differences in toxicity to *Ceriodaphnia*. Individuals exposed to static conditions slightly overestimated toxicity, while the flow-through renewal system clearly underestimated toxicity when compared with *in situ* survival data (i.e., static = 0 to 36.7% > *in situ* = 14.0 to 33% > flow-through = 76.7 to 83.3%).

3.5 Acceptability of freshwater sediment bioassays

Over the past few years, several reviews on bioassays for freshwater sediment assessment have been published (Giesy and Hoke, 1989; Burton, 1991, 1992a,b; Burton et al., 1992; Reynoldson and Day, 1993; Ingersoll, 1995), as have standardized protocols (ASTM, 1994). The reader should address these sources for comprehensive and detailed information. Detailed information on standard methods for bioassays in aqueous phases can be obtained from the extensive literature dealing with bioassessment of effluents and receiving waters or testing chemicals in aqueous media (e.g., ASTM, 1980; Organization for Economic Cooperation and Development, 1983; U.S. EPA, 1989, 1991; Canadian Council of Ministers of the Environment, 1993). Tables 3-6 to 3-8 summarize the bioassays that have been reported in the reviewed

Table 3-6 Sediment Microbiological Bioassays

Bioassay	Reference
Microtox®	Giesy et al., 1988a,b, 1990; Ross and Henebry, 1989; Ankley et al., 1990; Hoke et al., 1990, 1992b, 1993; Jacobs et al., 1993; Santiago et al., 1993
TOXI Chromotest® with *Escherichia coli* + alkaline phosphatase activity	Ahlf and Wild-Metzko, 1992
Microtox® and MetPAD (*E. coli, Photobacterium phosphoreum*)	Bitton et al., 1992
Cytotoxicity and enzymatic bioassays	Bitton and Koopman, 1992
Sediment-contact bioassay (*P. phosphoreum*)	Brouwer et al., 1990
α-D-Glucosidase inhibition (*Bacillus licheniformis*) and Microtox®	Campbell et al., 1993
Inhibition of luminescence and enzyme activity	Chen et al., 1994
Microtox® SPT and Chromotest® solid-phase tests	Day et al., 1995
Growth inhibition (*Colpidium campylum*, protozoa)	Dive et al., 1989
Mutagenicity and luminescence inhibition	Dutka and Kwan, 1988
Microtox® + SOS genotoxicity + ATP-TOX *Spirillum volutans* + ATP inhibition in *Selenastrum*	Dutka et al., 1989
SOS-Chromotest® (genotoxicity and cytotoxicity)	Dutka et al., 1995
Oxygen consumption (*Aeromonas hydrophila*)	Flemming and Trevors, 1989
Mutagenicity (*Salmonella typhimurium*)	Grifoll et al., 1990
CO_2 production (*E. coli*)	Jardim et al., 1990
Mutagenicity (MUTATOX)	Kwan et al., 1990; Johnson, 1992
Sediment Chromotest® (*E. coli*)	Kwan, 1995; Wong et al., 1994
Sediment ChromoPad and Microtox® SPT	Kwan and Dutka, 1995
Inhibition of respiration (activated muds, sediment)	Kwan, 1988, 1993
Enzymatic activity (*B. cereus*)	Liu, 1989
Metabolism: respiration, protein synthesis, enzyme activity, etc.	Reichardt et al., 1993

Table 3-7 Sediment Algae and Plant Bioassays

Algae bioassay	End point	Reference
Ankistrodesmus bibraianus with elutriates	72-hr growth inhibition	Ahlf et al., 1989; Ahlf and Wild-Metzko, 1992
Lolium multiflorum and *Lepidium sativum* with diluted sediments	Seed germination, plant weight production	Ahlf et al., 1989
Selenastrum capricornutum liquid phase	7-day growth inhibition	Ankley et al., 1990
S. capricornutum liquid phase	48-hr growth inhibition	Burton et al., 1989
Limited sediment bioassay	4-hr ^{14}C uptake	Gregor and Munawar, 1989; Munawar et al., 1989a
Algal fractionation bioassay with elutriates	4-hr ^{14}C uptake, chlorophyll *a*	Munawar et al., 1985; Munawar and Munawar, 1987; Gregor and Munawar, 1989; Santiago et al., 1993
Solid-phase and elutriate chronic tests, *Chlorella vulgaris*	<96-hr ^{14}C uptake	Munawar and Munawar, 1987
VAS-fluorescence assay, *A. braunii*	Chlorophyll fluorescence	Munawar et al., 1989c
S. capricornutum liquid phase	Photosynthesis inhibition, ^{14}C uptake	Ross et al., 1988; Ross and Henebry, 1989

literature published in recent years with information on the sediment phase and measured end points.

Any test organism has the potential to show a variable response to toxicants. This may be related to genetic differences, variations in life cycle that can affect individual biomass, or the general condition of the organism. Because of this inherent variability in organisms, it is essential that methods be developed to assess the performance of the test organisms and the validity of test results as part of standard laboratory procedures. This is usually carried out by establishing performance acceptability criteria for the test based on both positive and negative controls. Negative controls usually involve repeating the test conditions with either a culture medium or sediment. Positive controls involve the use of reference toxicants.

Responses in the control series should meet acceptability criteria for the appropriate end points measured, usually indicated in the standardized protocols. If these are unavailable, the researcher should develop an intralaboratory chart of the acceptable range of values for the end points (e.g., control charts) (U.S. EPA, 1989). These are based on routine laboratory performance of the bioassays, which consists of a cumulative mean value with the confidence limits for each end point and control sediment.

Reference toxicants, such as copper sulfate, dichromate, or phenol, have been widely used for measuring the health and sensitivity of organisms used in bioassays. This practice is recommended routinely with animals from the culture stock, or pairwise conducted with sediment bioassays, to establish the acceptable sensitivity of the test organisms. These are commonly conducted in the absence of sediment in acute tests, measuring percent mortality or immobilization. For *Hyalella azteca*, *Chironomus tentans*, and *Hexagenia limbata*, the use of an artificial substrate such as plastic mesh, sand, and glass tubes, respectively, is recommended to reduce the stress due to sediment absence (Burton and MacPherson, 1995). The development of standardized spiking procedures on artificial sediments would facilitate the intercalibration for positive controls using sediment bioassays.

The greatest source of variation among tests is due to age (or initial biomass) and the physiological and nutritional state of the test organisms (Giesy and Hoke, 1989). These variables are a common cause of failing acceptability criteria in bioassays. Therefore, these variables should be standardized in the test protocols as much as possible.

The quality and acceptability of results will also be affected by the physico-chemical conditions (such as pH, temperature, dissolved oxygen) under which the bioassay is conducted. These parameters should all be monitored during the test.

3.6 Expression of the results and statistical analysis

There are a number of approaches that can be used to express results of laboratory toxicity assessments. Basic issues of statistics of experimental design in toxicity tests and bioassays, including the number of replicates, concentrations, and animals per test chamber, are described by Hodson et al. (1977).

Organism responses in dose–response bioassays with whole sediment or their extracts can be expressed in the form of either the effective concentration (EC), at which commonly 50% of the exposed individuals show a statistically significant response, or the no observed effect concentration (NOEC) and the lowest observed effect concentration (LOEC). The EC_{50} can be calculated using various methods (for example, U.S. EPA, 1991; Newman, 1995), the Probit (Finney, 1971) being the most commonly used. However, other methods may also be applied to the data obtained depending on the experimental design and the number of intermediate responses (e.g., Litchfield and Wilcoxon, moving averages, geometric mean, trimmed Spearman–Karber tests, inverse regression, and others). The U.S. EPA has

Table 3-8 Sediment Animal Bioassays

Test species	Sediment phase	End point[a]	Reference
Nematoda			
Panagrellus redivivus	Sediment chemical fractions	96-hr survival, growth and maturation inhibition	Samoiloff et al., 1983; Gregor and Munawar, 1989
	Elutriate	96-hr lethality and molting inhibition	Ross and Henebry, 1989
Oligochaeta			
Stylodrilus heringianus and *Limnodrilus hoffmeisteri*	Endrin-spiked sediment	96-hr sediment avoidance, survival, species interaction, reworking rates, growth	Keilty et al., 1988a
S. heringianus	Microcosms	Survival, weight loss, sediment reworking rates	Keilty and Landrum, 1990
	Endrin-spiked sediment	Reworking rates, survival, bioaccumulation	Keilty et al., 1988c
S. heringianus and *Tubificidae* spp.	Solid phase	96-hr survival	Chapman et al., 1982
Tubificidae (several species)	Whole and Cu-spiked	500-day survival, growth, reproduction	Wiederholm et al., 1987
Monopylephorus cuticulatus	Elutriate	Respiration	Chapman, 1987
Branchiura sowerbyi	LAS-spiked sediment	220-day reproduction	Casellato et al., 1992
L. hoffmeisteri and *S. sowerbyi*	LAS dosed to water	96-hr survival and 3-month reproduction	Casellato and Negrisolo, 1989
Tubifex tubifex and *L. hoffmeisteri*	Solid phase	Pesticides, PCBs, bioaccumulation	Markwell et al., 1989
T. tubifex	Solid phase	Growth and reproduction	Milbrink, 1987
	Solid phase	Survival, reproduction	Reynoldson et al., 1991
	Solid phase	Survival, reproduction	Reynoldson, 1994
	Solid phase	Effect of indigenous benthos	Reynoldson et al., 1994

Lumbriculus variegatus	Solid phase	AVS, Cd and Ni bioavailability	Ankley et al., 1991a
	Solid phase	30-day bioaccumulation of PCBs	Ankley et al., 1992a
	Solid phase	10-day survival, growth, reproduction	Ankley et al., 1994a
	Aqueous and solid phase	96-hr survival	Ankley et al., 1991b
	Solid phase	Borrowing, 2-week survival, growth, reproduction	Dermott and Munawar, 1992
	Aqueous and solid phase	4- and 7-day accumulation of organics	Harkey et al., 1994
	Solid	10- to 60-day survival, growth, reproduction, bioaccumulation	Phipps et al., 1993
	Aqueous and solid phase	4-day survival	Schubauer-Berigan and Ankley, 1991
	Solid phase	10-day survival	West et al., 1993
Pristina leidyi	Aqueous and solid phase	Survival and population growth	Smith et al., 1991
Amphipoda			
Hyalella azteca	Solid phase	AVS, Cd and Ni bioavailability	Ankley et al., 1991a
	Pore water and solid phase	10-day survival	Ankley et al., 1991b
	Aqueous and solid phase	96-hr survival	Ankley et al., 1991b
	Solid phase	10-day survival	Ankley et al., 1994a
	Solid phase	Chronic bioaccumulation	Borgmann et al., 1991
	Solid phase	Survival, growth	Borgmann and Munawar, 1989; Borgmann and Norwood, 1993
	Elutriate and solid phase	48-hr survival	Burton et al., 1989
	Solid phase	Pesticide, EqP, survival	Hoke et al., 1994
	Spiked sediment	Dieldrin, EqP, 10-day survival	Hoke et al., 1995
	Solid phase	AVS predictions, Cd and Ni, 10-day survival	Di Toro et al., 1992
	Pore water	14-day survival, growth in Cu-spiked sediments	Kubitz et al., 1995

Table 3-8 Sediment Animal Bioassays (continued)

Test species	Sediment phase	End point[a]	Reference
Amphipoda (continued)			
Hyalella azteca (continued)	Solid phase	96-hr survival, Cd-spiked sediment	Nebeker et al., 1986
	Solid phase	DDT and endrin toxicity, survival	Nebeker et al., 1989
	Cd-spiked sediment	10-day survival	Nebeker et al., 1986
	Diluted sediments	14-day flow-through, survival	Nelson et al., 1993
	Solid phase	28-day survival and growth, indigenous animals	Reynoldson et al., 1994
	Aqueous and solid phase	48-hr survival	Schubauer-Berigan and Ankley, 1991
	Pore water	96-hr survival	Schubauer-Berigan et al., 1993
	Fluoranthene-spiked sediment	10-day survival	Suedel et al., 1993
	Artificial sediment	Chronic survival and reproduction	Suedel and Rodgers, 1994
	Solid phase	Effect of freezing	Schuytema et al., 1989
	Solid phase	10-day survival	West et al., 1993
	Pore water and solid phase	10-day static renewal, survival and leaf consumption	Winger and Lasier, 1995
Mulinia lateralis and *Ampelisca abdita*	Cu-spiked sand	?	Burgess et al., 1994
Diporeia sp.	Aqueous and solid phase	96-hr, 14-day bioaccumulation of organics	Harkey et al., 1994
	Solid phase	PAH toxicokinetics	Landrum et al., 1991
	Solid phase	EqP, pyrene and phenanthrene, LD_{50}, respiration, spiked sediment, bioaccumulation	Landrum et al., 1994
	Spiked sediment	PAH and PCBs, uptake clearance	Landrum and Faust, 1994
	Diluted sediment	14-day flow-through, survival	Nelson et al., 1993

Gammarus pulex	Solid phase	28-day survival and drift, influence of pH and Cu	Taylor et al., 1994
Isopoda			
Asellus aquaticus	Solid phase	Accumulation in microcosm	Van Hattum et al., 1993
Cladocera			
Daphnia magna, Ceriodaphnia dubia	Elutriate and solid phase	48-hr survival	Burton, 1989; Burton et al., 1989
	Pore water	48-hr survival	Hoke et al., 1992a, 1993
	Solid phase	48-hr survival	Stemmer et al., 1990b
	Artificial sediment	28-day survival	Suedel and Rodgers, 1994
C. dubia	Pore water and elutriates	48-hr survival	Ankley et al., 1991b; Schubauer-Berigan and Ankley, 1991
	Pore water	7-day reproduction test	Ankley et al., 1990
	Pore water	48-hr mortality	Ankley et al., 1992b
	Elutriates and solid phase	48-hr survival	Burton et al., 1989
	Pore water and elutriates	7-day survival and reproduction	Hoke et al., 1990
	Pore water	48-hr survival in Cu-spiked sediments	Kubitz et al., 1995
	Diluted sediments	7-day semi-static, survival, reproduction	Nelson et al., 1993
	Aqueous and solid phase	48-hr survival	Sasson-Brickson and Burton, 1991
	Pore water	48-hr survival	Schubauer-Berigan et al., 1993
D. magna	Elutriates	48-hr survival	Dutka et al., 1989
	Pore water	48-hr survival	Giesy et al., 1988b, 1990
	Solid phase	48-hr survival, 10-day reproduction	Othoudt et al., 1991
	Cd-spiked sediment	48-hr survival	Nebeker et al., 1986
	Pore water	Fluorescence, 48-hr survival in Cu-spiked sediments	Kubitz et al., 1995

Table 3-8 Sediment Animal Bioassays (continued)

Test species	Sediment phase	End point[a]	Reference
Cladocera (continued)			
D. magna (continued)	Cu-spiked sediment	48-hr survival, effect of storage	Malueg et al., 1986
	Elutriates and solid phase	7- to 10-day survival and reproduction	Nebeker et al., 1986
	Elutriates	24-hr survival	Santiago et al., 1993
	Fluoranthene-spiked sediment	10-day survival	Suedel et al., 1993
	Se-spiked sediment	48-hr survival	Stemmer et al., 1990a
D. similis	Aqueous extracts	24-hr EC_{50}	Zagatto et al., 1987
Ephemeroptera			
Stenonema modestum	?	Molting	Diamond et al., 1992
Hexagenia limbata	Pore water and solid phase	7-day survival, molting frequency, borrowing	Giesy et al., 1990
	Solid phase	21-day survival and growth	Reynoldson et al., 1994
Diptera, Chironomidae			
Chironomus tentans	Solid phase	TOC partitioning, pesticide	Ankley et al., 1994b
	Solid phase	10-day survival, growth	Ankley et al., 1994a
	Solid phase	10-day growth reduction	Giesy et al., 1988b
	Pore water and solid phase	10-day growth reduction	Giesy et al., 1990
	Pore water and solid phase	10-day growth inhibition	Hoke et al., 1990, 1993
	Dieldrin-spiked sediment	EqP, 10-day growth	Hoke et al., 1995
	Solid phase	25-day adult emergence, 15-day survival and growth	Nebeker et al., 1988
	Solid phase	14-day survival and growth	Othoudt et al., 1991
	Fluoranthene-spiked sediment	10-day survival	Suedel et al., 1993

	Artificial sediment	40-day survival, growth, development	Suedel and Rodgers, 1994
	Solid phase	10-day growth inhibition	Rosiu et al., 1989
	Solid phase	10-day survival	West et al., 1993, 1994
C. riparius	Solid phase	Pesticide, 24-hr acute test, effect of temperature and pH	Bruner and Fisher, 1993
	Solid phase	Sexual dimorphism on growth	Day et al., 1994
	Aqueous and solid phase	96-hr bioaccumulation of organics	Harkey et al., 1994
	Diluted sediments	14-day flow-through, survival	Nelson et al., 1993
	Solid phase	Effect of food and artificial sediment	Pascoe et al., 1990
	Solid phase	Survival, growth, deformities, emergence	Pellinen et al., 1993
	Solid phase	10-day survival and growth, indigenous animals	Reynoldson et al., 1994
Vertebrata			
Pimephales promelas	Pore water	96-hr survival	Ankley et al., 1990
Several fish and amphibian species	Field and spiked sediment	4 or 10 days after hatching survival, 20-day hatchability, teratogenesis, tissue metal concentration	Birge et al., 1987
P. promelas and frog larva	Aqueous extracts	7-day larval survival and growth	Dawson et al., 1988
P. promelas	Elutriates	7-day larval survival and growth	Hoke et al., 1990

[a] AVS = acid-volatile sulfur, EqP = equilibrium partitioning.

computer software that uses the Probit Analysis program for determining EC_{50} values with confidence limits and for determining the NOEC. The limit effect values LOEC and NOEC can be calculated by parametric ANOVA, provided that normality assumptions are filled, followed by post hoc tests such as Duncan's multiple range test or Dunnett *t*-tests. When tests of normality fail, the nonparametric Mann–Whitney U or Wilcoxon rank sum test can be used to compare each treatment to the control. The significance level of the analysis is commonly 0.05.

Correlation between the concentrations of the chemicals in the sediment or its extracts and the measured end points can be calculated. For nonpolar chemicals, the equilibrium partitioning theory states that the contaminants will preferably bind to organic carbon, with effects on organisms better correlated with normalized concentrations of organic carbon (e.g., fluoranthene) (Suedel et al., 1993). Ankley et al. (1993b) reviewed the basis of the prediction of bioavailability of cationic metals to benthic organisms based on the relationship between the concentrations of metals in sediment pore water and that of acid-volatile sulfur. The limitation of this approach is that it analyzes the effect of the chemicals individually, when the effect is the result of a global combination of chemicals and physical conditions in the sediment. A more appropriate method for identifying dose–response patterns is relating these multivariately, using either discriminant analysis or principal components correlation.

The contribution of metals or organics to toxicity can be estimated using the toxic units (TUs) calculated by dividing the concentration of metal in the sediment (i.e., in whole sediment or pore water) by its respective experimentally derived EC_{50} value and by contrasting the total TUs due to chemical values with the measured TU of each sample (Dave, 1992a,b; Hoke et al., 1993). This approach is inconvenient in that the concentrations of many chemicals in the sediment are not quantitatively measured, and the mixture of chemicals can produce antagonistic or synergistic effects which are difficult to predict.

Chapman (1990) normalized test results using the ratio to reference criteria in which individual measurements are divided by the same end point value from a reference sediment. The normalization procedure can also be applied to chemical and benthic community data, and each group of data can be combined as a single mean value for toxicity, chemistry, and field data in the TRIAD (Chapman, 1990) approach. Therefore, in this approach, the characteristics of the reference sediment(s) are important in interpretation of the results, and the selection criteria should be critically evaluated.

3.7 *Laboratory bioaccumulation tests*

Bioaccumulation is the uptake and sequestration of material from water and sediment by organisms (Phillips, 1993). General approaches to modeling bioaccumulation tests are addressed by Newman (1995). When chemicals are known to bioaccumulate, examination of animal tissues may be performed. However, this should never be used as a screening or classification tool but rather as a risk assessment of sediment-bound chemicals. Most bioaccumulation studies are performed using caged field organisms exposed to *in situ* concentrations of contaminants (see Chapter 4 of this book). It is also possible to measure the accumulation of chemicals by test organisms exposed to polluted sediments under laboratory conditions. A protocol has been proposed by Mac et al. (1990) for measuring the accumulation of contaminants in test organisms exposed in laboratory bioassays. The fathead minnow (*Pimephales promelas*) is recommended by these authors in a 10-day exposure to whole sediments in flow-through conditions. Some bioaccumulation studies have recently been conducted with invertebrates (Ingersoll et al., 1994; Landrum and Faust, 1994; Landrum et al., 1994). Bioaccumulation rates with mixtures of species in densities close to natural populations have been studied in a microcosm test system for aquatic oligochaetes (Keilty et al., 1988a) and isopods (Van Hattum et al., 1993).

3.8 Discussion and recommendations

In recent years, an increasing number of publications have proposed standard protocols for sediment bioassessment, including tiered approaches from biochemical/cellular to community data. Laboratory toxicity tests are one of the tools for conducting sediment bioassessments, and considerable work is still required to standardize, intercalibrate, and validate laboratory bioassays. However, they have already been successfully incorporated in screening and remediation programs and the development of biological quality criteria for sediments in the North American Great Lakes (Reynoldson et al., 1995).

Aquatic sediments are complex, hierarchical systems whose toxic behaviors are difficult to predict, with numerous variables interacting in laboratory bioassays. There is no "cookbook" approach to selecting the "best test," as such a thing does not exist. However, the authors can suggest the following general recommendations for laboratory bioassays in a sediment bioassessment program:

1. Use a battery of species, documented as sensitive, at least some of which are sediment-associated species. The effects on different organisms at various trophic levels should be considered, as biological impairment may occur by different routes and at different degrees in the ecosystem.
2. Preferably expose organisms to bulk sediment instead of sediment extracts, although these can be useful for assessing impacts of dredging or comparing the sensitivity of species.

3. Combine acute and chronic bioassays, in a tiered approach, to build a classification of sediments from highly polluted to clean sites that facilitates intervention priorities.
4. Choose several end points, from early biochemical or cellular measures to organism-level responses and eventually to the multispecies level, for assessing the fate of chemicals.
5. Utilize the results from each technique to reduce subsequent sampling requirements and, therefore, costs and to avoid redundant information.
6. Intercalibrate or create intralaboratory control charts to establish the acceptability of the bioassays, using both a control sediment series and reference toxicants.
7. Use reference sediments to predict which effects can be expected depending on sediment characteristics other than the concentration of contaminants.
8. Integrate field benthic community data along with physico-chemical and toxicity data, to provide an accurate assessment of specific problems in contaminated sediment. The data should provide evidence of the linkage between contaminated sediment and the problem (i.e., cause–effect relationship).

Once the results have been obtained, the following questions should be considered to evaluate the methods being used:

1. Do the methods provide adequate proof of the linkage between the contaminated sediments and the observed effect (i.e., cause–effect relationship)?
2. Do they quantify problem severity, thereby enabling intercomparisons between and within areas of investigation?

Most of the sites that are the subject of bioassessment surveys are not "clean" or "heavily polluted," but are rated intermediate (i.e., with respect to the concentration of contaminants). These so-called "gray" sites (Chapman, 1989) represent the most common condition in sediment in many urban areas, and classification of these sites is by far the most complex, as it is not rare to find contradictory results with different bioassays. A stimulatory effect at low levels of contaminant concentrations, known as "hormesis" (Laughlin et al., 1981), has been described in microbial community, algae, macrophyte, and animal bioassays (Burton et al., 1992). Enhancement may be a first sign of stress preceding a final observation of severe impoverishment (Cairns et al., 1992). Thus, both enhancement and inhibition in the response compared with a control or a reference site can be interpreted under certain circumstances as perturbation due to pollution. Therefore, the use of the "weight of evidence" approach (Chapman, 1995a) is required to evaluate sediment quality.

The need for validation of laboratory bioassay data in the field has been a concern of scientists and managers for many years. Thus, the increasing number of protocols for conducting bioassays with bulk sediment, using sediment-associated species, and measuring effects under chronic exposures are more ecologically relevant. The search for trends of cause and effect with statistical significance using multivariate analysis that integrates toxicity responses, sediment chemistry, and benthic community structure data will allow a degree of prediction of field conditions. Laboratory and field studies are different approaches to the same problem, each providing different information and together building an overall perspective for sediment bioassessment (Chapman, 1995b).

References

Adams, D.D., Sampling sediment pore water, in *Handbook of Techniques for Aquatic Sediments Sampling*, Mudroch, A. and MacKnight, S.D., Eds., CRC Press, Boca Raton, FL, 1991, 171.

Ahlf, W. and Wild-Metzko, S., Bioassay responses to sediment elutriates and multivariate data analysis for hazard assessment of sediment-bound chemicals, *Hydrobiologia*, 235/236, 415, 1992.

Ahlf, W., Calmano, W., Erhard, J., and Förstner, U., Comparison of five bioassay techniques for assessing sediment-bound contaminants, *Hydrobiologia*, 188/189, 285, 1989.

American Society for Testing and Materials, Standard practice for conducting acute toxicity tests with fishes, macroinvertebrates and amphibians, *Annual Book of ASTM Standards*, 11, 272, 1980.

American Society for Testing and Materials, *Standard Guide for Collection, Storage, Characterization, and Manipulation of Sediments for Toxicological Testing*, ASTM, Philadelphia, 1990.

American Society for Testing and Materials, *Standard Guide for Conducting Sediment Toxicity Tests with Freshwater Invertebrates*, ASTM, Philadelphia, 1994.

Ankley, G.T., Katko, A., and Arthur, J., Identification of ammonia as an important sediment-associated toxicant in the Lower Fox River and Green Bay, Wisconsin, *Environ. Toxicol. Chem.*, 9, 313, 1990.

Ankley, G.T., Phipps, G.L., Leonard, E.N., Benoit, B.D.A., Mattson, V.R., Kosian, P.A., Cotter, A.M., Dierhes, J.R., Hansen, D.J., and Mahony, J.D., Acid-volatile sulfide as a factor mediating cadmium and nickel bioavailability in contaminated sediments, *Environ. Toxicol. Chem.*, 10, 1299, 1991a.

Ankley, G.T., Schubauer-Berigan, M.K., and Dierkes, J.R., Predicting the toxicity of bulk sediments to aquatic organisms with aqueous test fractions — pore water vs. elutriate, *Environ. Toxicol. Chem.*, 10, 1359, 1991b.

Ankley, G.T., Cook, P.M., Carlson, A.R., Call, D.J., Swenson, J.A., Corcoran, H.F., and Hoke, R.A., Bioaccumulation of PCBs from sediments by oligochaetes and fish, comparison of laboratory and field studies, *Can. J. Fish. Aquat. Sci.*, 49, 2080, 1992a.

Ankley, G.T., Lodge, K., Call, D.J., Balcer, M.D., Brooke, L.T., Cook, P.M., Kreis, R.G., Carlson, A.R., Johnson, R.D., Niemi, G.J., Hoke, R.A., West, C.W., Giesy, J.P.,

Jones, P.D., and Fuying, Z.C., Integrated assessment of contaminated sediments in the Lower Fox River and Green Bay, Wisconsin, *Ecotoxicol. Environ. Saf.*, 23, 46, 1992b.

Ankley, G.T., Mattson, V.R., Leonard, E.N., West, C.W., and Bennett, J.L., Predicting the acute toxicity of copper in freshwater sediments: evaluation of the role of acid-volatile sulfide, *Environ. Toxicol. Chem.*, 12, 315, 1993a.

Ankley, G.T., Benoit, B.D.A., Hoke, R.A., Leonard, E.N., West, C.W., Phipps, G.L., Mattson, V.R., and Anderson, L.A., Development and evaluation of the test methods for benthic invertebrates and sediments: effects of flow-rate and feeding on water quality and exposure conditions, *Arch. Environ. Contam. Toxicol.*, 25, 12, 1993b.

Ankley, G.T., Benoit, B.D.A., Balogh, J.C., Reynoldson, T.B., Day, K.E., and Hoke, R.A., Evaluation of potential confounding factors in sediment toxicity tests with three freshwater benthic invertebrates, *Environ. Toxicol. Chem.*, 13, 627, 1994a.

Ankley, G.T., Call, D.J., Cox, J.S., Kahl, M.D., Hoke, R.A., and Kosian, P.A., Organic carbon partitioning as a basis for predicting the toxicity of chlorpyrifos in sediments, *Environ. Toxicol. Chem.*, 13, 621, 1994b.

Becker, D.S. and Ginn, T.C., Effects of storage time on toxicity of sediments from Puget Sound, Washington, *Environ. Toxicol. Chem.*, 14, 829, 1995.

Benoit, D.A., Philipps, G.L., and Ankley, G.T., A sediment testing intermittent renewal system for the automated renewal of overlying water in toxicity tests with contaminated sediments, *Water Res.*, 27, 1403, 1993.

Birge, W.J., Black, J.A., Westerman, A.G., and Francis, P., Toxicity of sediment-associated metals to freshwater organisms: biomonitoring processes, in *Fate and Effects of Sediment-Bound Chemicals in Aquatic Systems*, Dickson, K.L., Maki, A.W., and Brungs, W.A., Eds., SETAC–Special Publ. Ser., Pergamon Press, New York, 1987, 199.

Bitton, G. and Koopman, B., Bacterial and enzymatic bioassays for toxicity testing in the environment, *Rev. Environ. Contam. Toxicol.*, 125, 1, 1992.

Bitton, G., Campbell, M., and Koopman, B., MetPAD: a bioassay kit for the specific determination of heavy metal toxicity in sediments from hazardous waste sites, *Environ. Toxicol. Water Qual.*, 7, 3223, 1992.

Bode, R.W. and Novak, M.A., Development and application of biological criteria for rivers and streams in New York State, in *Biological Assessment and Criteria: Tools for Water Resource Planning and Decision Making*, Davis, W.S. and Simon, T.P., Eds., Lewis Publishers, Boca Raton, FL, 1995, 97.

Borgmann, U. and Munawar, M., A new standardized sediment bioassay protocol using the amphipod *Hyalella azteca* (Saussure), *Hydrobiologia*, 188/189, 425, 1989.

Borgmann, U. and Norwood, W.P., Spatial and temporal variability in toxicity of Hamilton Harbour sediments — evaluation of the *Hyalella azteca* 4 week chronic toxicity test, *J. Great Lakes Res.*, 19(1), 72, 1993.

Borgmann, U., Norwood, W.P., and Babirad, I.M., Relationship between toxicity and bioaccumulation of cadmium in *Hyalella azteca, Can. J. Fish. Aquat. Sci.*, 48, 1055, 1991.

Brouwer, H., Murphy, T.P., and McArdle, L., A sediment-contact bioassay with *Photobacterium phosphoreum, Environ. Toxicol. Chem.*, 9, 1353, 1990.

Bruner, K.A. and Fisher, S.W., The effects of temperature, pH, and sediment on the fate and toxicity of 1-naphthol to the midge larvae *Chironomus riparius*, *J. Environ. Sci. Health*, 28, 1341, 1993.

Bufflap, S.E. and Allen, H.E., Sediment pore water collection methods for trace metal analysis: a review, *Water Res.*, 29, 165, 1995.

Buikema, A.L. and Voshell, J.R., Toxicity studies using freshwater benthic macroinvertebrates, in *Freshwater Biomonitoring and Benthic Macroinvertebrates*, Rosenberg, D.M. and Resh, V.H., Eds., Chapman and Hall, New York, 1993, 344.

Burgess, R.M., Rogers, B.A., Rego, S.A., Corbin, J.M., and Morrison, G.E., Sand spiked with copper as a reference toxicant material for sediment toxicity testing: a preliminary evaluation, *Arch. Environ. Contam. Toxicol.*, 26, 163, 1994.

Burton, G.A., Jr., Evaluation of seven sediment toxicity tests and their relationships to stream parameters, *Toxicol. Assess.*, 4, 149, 1989.

Burton, G.A, Jr., Assessing the toxicity of freshwater sediments, *Environ. Toxicol. Chem.*, 10, 1585, 1991.

Burton, G.A., Jr., Sediment collection and processing: factors affecting realism, in *Sediment Toxicity Bioassessment*, Burton, G.A., Jr., Ed., Lewis Publishers, Boca Raton, FL, 1992a, 37.

Burton, G.A., Jr., Plankton, macrophyte, fish, and amphibian toxicity testing of freshwater sediments, in *Sediment Toxicity Bioassessment*, Burton, G.A., Jr., Ed., Lewis Publishers, Boca Raton, FL, 1992b, 167.

Burton, G.A., Jr. and MacPherson, C., Sediment toxicity testing issues and methods, in *Handbook of Ecotoxicology*, Hoffman, D.J., Rattner, B.A., Burton, G.A., Jr., and Cairns, J., Jr., Eds., Lewis Publishers, Boca Raton, FL, 1995, 70.

Burton, G.A., Jr., Lazorchak, J.M., Waller, W.T., and Lanza, G.R., Arsenic toxicity changes in the presence of sediment, *Bull. Environ. Contam. Toxicol.*, 38, 491, 1987.

Burton, G.A., Jr., Stemmer, B.L., Winks, K.L., Ross, E.R., and Burnett, L.C., A multitrophic level evaluation of sediment toxicity in Waukegan and Indiana Harbors, *Environ. Toxicol. Chem.*, 8, 1057, 1989.

Burton, G.A., Jr., Nelson, M.K., and Ingersoll, C.G., Freshwater benthic toxicity tests, in *Sediment Toxicity Bioassessment*, Burton, G.A., Jr., Ed., Lewis Publishers, Boca Raton, FL, 1992, 213.

Cairns, J., Jr., Ed., *Multispecies Toxicity Testing*, A Special Publication of SETAC, Pergamon Press, New York, 1985.

Cairns, J., Jr. and Cherry, D.S., Freshwater multispecies test systems, in *Handbook of Ecotoxicology*, Calow, P., Ed., Blackwell Scientific, Oxford, 1993, 101.

Cairns, J., Jr. and Pratt, J.R., Trends in ecotoxicology, *Sci. Total Environ.*, Suppl. 1, 7, 1993.

Cairns, J., Jr., Kuhn, D.L., and Plafkin, J.L., Protozoa colonization of artificial substrates, in *Methods and Measurement of Periphyton Communities: A Review*, Wetzel, R.L., Ed., American Society for Testing and Materials, Philadelphia, 1979.

Cairns, J., Jr., Niederlehner, B.R., and Smith, E.P., The emergence of functional attributes as endpoints in ecotoxicology, in *Sediment Toxicity Bioassessment*, Burton, G.A., Jr., Ed., Lewis Publishers, Boca Raton, FL, 1992, 111.

Calow, P., The choice and implementation of environmental bioassays, *Hydrobiologia*, 188/189, 61, 1989.

Campbell, M., Bitton, G., and Koopman, B., Toxicity testing of sediment elutriates based on inhibition of alpha-glucosidase biosynthesis in *Bacillus licheniformis*, *Arch. Environ. Contam. Toxicol.*, 24, 469, 1993.

Canadian Council of Ministers of the Environment, Guidance Manual on Sampling, Analysis, and Data Management for Contaminated Sites, The National Contaminated Sites Remediation Program, 1993.

Casellato, S. and Negrisolo, P.A., Acute and chronic effects of an anionic surfactant on some freshwater tubificid species, *Hydrobiologia*, 180, 243, 1989.

Casellato, S., Aiello, R., Negrisolo, P.A., and Seno, M., Long-term experiment on *Branchiura sowerby* Beddard (Oligochaeta, Tubificidae) using sediment treated with LAS (linear alkylbenzene sulphonate), *Hydrobiologia*, 232, 169, 1992.

Chapman, P.M., Oligochaete respiration as a measure of sediment toxicity in Puget Sound, Washington, *Hydrobiologia*, 155, 249, 1987.

Chapman, P.M., Current approaches to developing sediment quality criteria, *Environ. Toxicol. Chem.*, 8, 589, 1989.

Chapman, P.M., The sediment quality triad approach to determining pollution-induced degradation, *Total Environ.*, 97/98, 815, 1990.

Chapman, P.M., Criteria. What type should we be developing? *Environ. Sci. Technol.*, 25, 1353, 1991.

Chapman, P.M., Extrapolating laboratory toxicity results to the field, *Environ. Toxicol. Chem.*, 14, 927, 1995a.

Chapman, P.M., Do sediment toxicity tests require field validation? *Environ. Toxicol. Chem.*, 14, 1451, 1995b.

Chapman, P.M., Farrell, M.A., and Brinkhurst, R.O., Relative tolerances of selected aquatic oligochaetes to individual pollutants and environmental factors, *Aquat. Toxicol.*, 2, 47, 1982.

Chapman, P.M., Power, E.A., and Burton, G.A., Jr., Integrative assessments in aquatic ecosystems, in *Sediment Toxicity Bioassessment*, Burton, G.A., Jr., Ed., Lewis Publishers, Boca Raton, FL, 1992, 313.

Chen, W., Morrison, G.M., Hamilton, R.S., Revitt, D.M., Harrison, R.M., and Monzon de Caceres, A., Inhibition of bacterial enzyme activity and luminescence by urban river sediments, *Total Environ.*, 146–147, 141, 1994.

Chilton, E.W., System for maintaining sediment suspensions during larval fish studies, *Prog. Fish Cult.*, 53, 28, 1991.

Coler, R.A., Coler, M.S., and Kostecki, P.T., Tubificid behavior as a stress indicator, *Water Res.*, 22, 263, 1988.

Daniels, S.A., Munawar, M., and Mayfield, C.I., An improved elutriation technique for the bioassessment of sediment contaminants, *Hydrobiologia*, 188/189, 619, 1989.

Dave, G., Sediment toxicity and heavy metals in eleven lime reference lakes of Sweden, *Water Air Soil Pollut.*, 63, 187, 1992a.

Dave, G., Sediment toxicity in lakes along the river Kolbäcksan, Central Sweden, *Hydrobiologia*, 235/236, 419, 1992b.

Dawson, D.A., Stebler, E.F., Burks, S.L., and Bantle, J.A., Evaluation of the developmental toxicity of metal-contaminated sediments using short-term fathead minnow and frog embryo-larval assays, *Environ. Toxicol. Chem.*, 7, 23, 1988.

Day, K.E., Kirby, R.S., and Reynoldson, T.B., Sexual dimorphism in *Chironomus riparius* (Meigen): impact on interpretation of growth in whole sediment toxicity tests, *Environ. Toxicol. Chem.*, 13, 35, 1994.

Day, K.E., Dutka, B.J., Kwan, K.K., Batista, N., Reynoldson, T.B., and Metcalfe-Smith, J.L., Correlations between solid-phase microbial screening assays, whole-sediment toxicity tests with macroinvertebrates and *in situ* benthic community structure, *J. Great Lakes Res.*, 21, 192, 1995.

Dermott, R. and Munawar, M., A simple and sensitive assay for evaluation of sediment toxicity using *Lumbriculus variegatus* (Müller), *Hydrobiologia*, 235/236, 407, 1992.

Diamond, J.M., Winchester, E.L., Mackler, D.G., and Gruber, D., Use of the mayfly *Stenonema modestum* (Heptageniidae) in subacute toxicity assessment, *Environ. Toxicol. Chem.*, 11, 415, 1992.

Di Toro, D.M., Mahony, J.D., Hansen, D.J., Scott, K.J., Carlson, A.R., and Ankley, G.T., Acid volatile sulfide predicts the acute toxicity of cadmium and nickel in sediments, *Environ. Sci. Technol.*, 26, 96, 1992.

Ditsworth, G.R., Schults, D.W., and Jones, J.K.P., Preparation of benthic substances for sediment toxicity testing, *Environ. Toxicol. Chem.*, 9, 1523, 1990.

Dive, D., Robert, S., Agrannd, E., Bel, C., Bonnemain, H., Brun, L., Demarque, Y., Le Du, A., El Bouhouti, R., Fourmaux, M.N., Guery, L., Hanssens, O., and Murat, M., A bioassay using the measurement of the growth inhibition of a ciliate protozoan: *Colpidium campylum* Stokes, *Hydrobiologia*, 188/189, 181, 1989.

Dutka, B.J. and Kwan, K.K., Battery of screening tests approach applied to sediment extracts, *Toxicol. Assess.*, 3, 303, 1988.

Dutka, B.J., Touminen, T., Churchland, L., and Kwan, K.K., Fraser River sediments and waters evaluated by the battery of screening tests technique, *Hydrobiologia*, 188/189, 301, 1989.

Dutka, B.J., Teichgräber, K., and Lifshitz, R., A modified SOS-Chromotest procedure to test for genotoxicity and cytotoxicity in sediments directly without extraction, *Chemosphere*, 31, 3273, 1995.

Egeler, P., Römbke, J., Knacker, T., and Dohalnass, R.N., Bioaccumulation of hexachlorobenzene by tubificid sludgeworms (oligochaeta) from spiked artificial sediments, Communication in 5th SETAC–Europe Congress, Copenhagen, June 25 to 28, 1995.

Elliott, J.M. and Tullet, P.A., *A Bibliography of Samplers for Benthic Invertebrates,* Freshwater Biological Association, Occasional Publ. 4, 1978, 1.

Faith, D.P. and Norris, R.H., Correlation of environmental variables with patterns of distribution and abundance of common and rare freshwater macroinvertebrates, *Biol. Conserv.*, 50, 77, 1989.

Finney, D.J., *Probit Analysis*, 3rd ed., Cambridge Press, New York, 1971, 668.

Flemming, C.A. and Trevors, J.T., Copper toxicity in freshwater sediment and *Aeromonas hydrophila* cell suspensions measured using an O_2 electrode, *Toxicol. Assess.*, 4, 473, 1989.

Giesy, J. and Hoke, R., Freshwater sediment toxicity bioassessment: rationale for species selection and test design, *J. Great Lakes Res.*, 15, 539, 1989.

Giesy, J.P., Graney, R.L., Newsted, J.L., Rosiu, C.J., Benda, A., Kreis, R.G., Jr., and Horvath, F.J., Comparison of three sediment bioassay methods using Detroit River sediments, *Environ. Toxicol. Chem.*, 7, 483, 1988a.

Giesy, J.P., Rosiu, C.J., Graney, R.L., Newsted, J.L., Benda, A., Kreis, R.G., Jr., and Horvath, F.J., Toxicity of Detroit River sediments interstitial water to the bacterium *Photobacterium phosphoreum*, *J. Great Lakes Res.*, 14, 502, 1988b.

Giesy, J.P., Rosiu, C.J., Graney, R.L., and Henry, M.G., Benthic invertebrate bioassays with toxic sediment and pore water, *Environ. Toxicol. Chem.*, 9, 233, 1990.

Gregor, D.J. and Munawar, M., Assessing toxicity of Lake Diefenbacker (Saskatchewan, Canada) sediments using algal and nematode bioassays, *Hydrobiologia*, 188/189, 291, 1989.

Grifoll, M., Solana, A.M., and Bayona, J.M., Characterization of genotoxic components in sediments by mass spectrometric techniques combined with Salmonella/microsome test, *Arch. Environ. Contam. Toxicol.*, 19, 175, 1990.

Håkanson, L., Sediment variability, in *Sediment Toxicity Bioassessment*, Burton, G.A., Jr., Ed., Lewis Publishers, Boca Raton, FL, 1992, 19.

Harkey, G.A., Landrum, P.F., and Klaine, S.J., Comparison of whole-sediment, elutriate and pore-water exposures for use in assessing sediment-associated organic contaminants in bioassays, *Environ. Toxicol. Chem.*, 13, 1315, 1994.

Hill, I.R., Matthiessen, P., and Heimbach, F., Eds., Guidance document on sediment toxicity tests and bioassays for freshwater and marine environments, in SETAC–Europe Workshop on Sediment Toxicity Assessment, Slot Moermond Congrescentrum Renesse, Netherlands, November 8 to 10, 1993.

Hodson, P.V., Ross, C.W., Niimi, A.J., and Spry, D.J., Statistical Considerations in Planning Aquatic Bioassays, Surveillance Report EPS-5-AR-77-1, Environment Canada, Halifax, 1977, 15.

Hoke, R.A. and Prater, B.L., Relationship of per cent mortality of four species of aquatic biota from 96-hour sediment bioassay of five Lake Michigan harbors and elutriate chemistry of the sediment, *Bull. Environ. Contam. Toxicol.*, 25, 394, 1980.

Hoke, R.A., Giesy, J.P., Ankley, G.T., Newsted, J.L., and Adams, J.R., Toxicity of sediments from western Lake Erie and the Maumee River at Toledo, Ohio, 1987: implications for current dredged material disposal practices, *J. Great Lakes Res.*, 16(3), 457, 1990.

Hoke, R.A., Gala, W.R., Drake, J.B., and Giesy, J.P., Bicarbonate as a potential confounding factor in cladoceran toxicity assessments of pore water from contaminated sediments, *Can. J. Fish. Aquat. Sci.*, 49, 1633, 1992a.

Hoke, R.A., Giesy, J.P., and Kreis, R.G., Sediment pore water toxicity identification in the Lower Fox River and Green Bay, Wisconsin, using the Microtox assay, *Ecotoxicol. Environ. Saf.*, 23, 343, 1992b.

Hoke, R.A., Giesy, J.P., Zabik, M., and Unger, M., Toxicity of sediments and sediment pore water from the Grand Calumet River–Indiana Harbour, Indiana Area of Concern, *Ecotoxicol. Environ. Saf.*, 26, 86, 1993.

Hoke, R.A., Ankley, G.T., Cotter, A.M., Goldstein, T., Kosian, P.A., Phipps, G.L., and Vandermeiden, F.M., Evaluation of equilibrium partitioning theory for predicting acute toxicity of field-collected sediments contaminated with DDT, DDE, and DDD to the amphipod *Hyalella azteca*, *Environ. Toxicol. Chem.*, 13, 157, 1994.

Hoke, R.A., Kosian, P.A., Ankley, G.T., Cotter, A.M., Vandermeiden, F.M., Phipps, G.L., and Durhan, E.J., Check studies with *Hyalella azteca* and *Chironomus tentans* in support of the development of a sediment quality criterion for dieldrin, *Environ. Toxicol. Chem.*, 14, 435, 1995.

Hughes, R.M., Defining acceptable biological status by comparing with reference conditions, in *Biological Assessment and Criteria: Tools for Water Resource Plan-*

ning and Decision Making, Davis, W.S. and Simon, T.P., Eds., Lewis Publishers, Boca Raton, FL, 1994, 31.

Hynes, H.B.N., *The Biology of Polluted Waters*, Liverpool University Press, Liverpool, U.K., 1960, 202.

Ingersoll, C.G., Sediment tests, in *Fundamentals of Aquatic Toxicology: Effects, Environmental Fate, and Risk Assessment*, 2nd ed., Rand, G.M., Ed., Taylor & Francis, Washington, D.C., 1995, 231.

Ingersoll, C.G., Brumbaugh, W.G., Dwyer, F.J., and Kemble, N.E., Bioaccumulation of metals by *Hyalella azteca* exposed to contaminated sediments from the upper Clark Fork River, Montana, *Environ. Toxicol. Chem.*, 13, 2013, 1994.

Ingersoll, C.G., Ankley, G.T., Benoit, D.A., Brunson, E.L., Burton, G.A., Dwyer, F.J., Hoke, R.A., Landrum, P.F., Norberg-King, T.J., and Winger, P.V., Toxicity and bioaccumulation of sediment-associated contaminants using freshwater invertebrates: a review of methods and applications, *Environ. Toxicol. Chem.*, 14, 1885, 1995.

Jacobs, M.W., Coates, J.A., Delfino, J.J., Bitton, G., Davis, W.M., and Garcia, K.L., Comparison of sediment extract Microtox® toxicity with semi-volatile organic priority pollutants concentrations, *Arch. Environ. Contam. Toxicol.*, 24, 461, 1993.

Jardim, W.F., Pasquini, C., Guimaraes, J.R., and De Faria, L.C., Short-term toxicity test using *Escherichia coli*: monitoring CO_2 production by flow injection analysis, *Water Res.*, 24, 351, 1990.

Johnson, B.T., Potential genotoxicity of sediments from the Great Lakes, *Environ. Toxicol. Water Qual.*, 7, 373, 1992.

Keilty, T.J. and Landrum, P.F., Population-specific toxicity responses by the freshwater oligochaete, *Stylodrilus heringianus*, in natural Lake Michigan sediments, *Environ. Toxicol. Chem.*, 9, 1147, 1990.

Keilty, T.J., White, D.S., and Landrum, P.F., Short-term lethality and sediment avoidance assays with endrin-contaminated sediment and two oligochaetes from Lake Michigan, *Arch. Environ. Contam. Toxicol.*, 17, 95, 1988a.

Keilty, T.J., White, D.S., and Landrum, P.F., Sublethal responses to endrin in sediment by *Limnodrilus hoffmeisteri* (Tubificidae), and in mixed culture with *Stylodrilus heringianus* (Lumbriculidae), *Aquat. Toxicol.*, 13, 227, 1988b.

Keilty, T.J., White, D.S., and Landrum, P.F., Sublethal responses to endrin in sediment by *Stylodrilus heringianus* (Lumbriculidae) as measured by a ^{137}cesium marker layer technique, *Aquat. Toxicol.*, 13, 251, 1988c.

Kennedy, J.H., Johnson, Z.B., Wise, P.D., and Johnson, P.C., Model aquatic ecosystems in ecotoxicological research: considerations of design, implementation, and analysis, in *Handbook of Ecotoxicology*, Hoffman, D.J., Rattner, B.A., Burton, G.A., Jr., and Cairns, J., Jr., Eds., Lewis Publishers, Boca Raton, FL, 1995, 117.

Kubitz, J.A., Leweke, E.C., Besser, J.M., Drake, J.B., III, and Giesy, J.P., Effects of copper-contaminated sediments on *Hyalella azteca*, *Daphnia magna*, and *Ceriodaphnia dubia*: survival, growth, and enzyme inhibition, *Arch. Environ. Contam. Toxicol.*, 29, 97, 1995.

Kwan, K.K., Synthetic activated sludge technique for toxicity assessment of chemical and environmental samples, *Toxicol. Assess.*, 3, 93, 1988.

Kwan, K.K., Direct solid phase toxicity testing procedure, *Environ. Toxicol. Water Qual.*, 8, 345, 1993.

Kwan, K.K., Direct sediment toxicity testing procedure using sediment Chromotest kit, *Environ. Toxicol. Water Qual.*, 9, 193, 1995.

Kwan, K.K. and Dutka, B.J., Comparative assessment of two solid-phase toxicity bioassays: the Direct Sediment Toxicity Testing Procedure (DSTTP) and the Microtox Solid-Phase Test (SPT), *Bull. Environ. Contam. Toxicol.*, 55, 338, 1995.

Kwan, K.K., Dutka, B.J., Rao, S.S., and Liu, D., Mutatox test: a new test for monitoring environmental genotoxic agents, *Environ. Pollut.*, 65, 323, 1990.

Landrum, P.F. and Faust, W.R., The role of sediment composition on the bioavailability of laboratory-dosed sediment-associated organic contaminants to the amphipod, *Diporeia* (spp.), *Chem. Spec. Bioavail.*, 6, 85, 1994.

Landrum, P.F., Eadie, B.J., and Faust, W.R., The role of sediment composition on the bioavailability of laboratory-dosed sediment-associated organic contaminants to the amphipod, *Diporeia* spp., *Environ. Toxicol. Chem.*, 10, 35, 1991.

Landrum, P.F., Dupuis, W.S., and Kukkonen, J., Toxicokinetics and toxicity of sediment-associated pyrene and phenanthrene in *Diporeia* (spp.): examination of equilibrium-partitioning theory and residue-based effects for assessing hazard, *Environ. Toxicol. Chem.*, 13, 1769, 1994.

Laughlin, R.B., Jr., Ng, J., and Guard, N.E., Hormesis: a response to low environmental concentration of petroleum hydrocarbons, *Science*, 211, 705, 1981.

Liu, D., A rapid and simple biochemical test for direct determination of chemical toxicity, *Toxicol. Assess.*, 4, 399, 1989.

Long, E.R. and Chapman, P.M., A sediment quality triad: measures of sediment contamination, toxicity and infaunal community composition in Puget Sound, *Mar. Pollut. Bull.*, 16, 405, 1985.

Luoma, S.N. and Carter, J.L., Understanding the toxicity of contaminants in sediments — beyond the bioassay based paradigm, *Environ. Toxicol. Chem.*, 12, 793, 1993.

Mac, M.J., Noguchi, G.E., Hesselberg, R.J., Edsall, C.C., Shoesmith, J.A., and Bowker, J.D., A bioaccumulation bioassay for freshwater sediments, *Environ. Toxicol. Chem.*, 9, 1405, 1990.

MacKnight, S.D., Selection of bottom sediment sampling stations, in *Handbook of Techniques for Aquatic Sediments Sampling*, Mudroch, A. and MacKnight, S.D., Eds., CRC Press, Boca Raton, FL, 1994, 17.

Malueg, K.W., Schuytema, G.S., Gakstatter, J.H., and Krawczyk, D.F., Effect of *Hexagenia* on *Daphnia* response in sediment toxicity tests, *Environ. Toxicol. Chem.*, 2, 73, 1983.

Malueg, K.W., Schuytema, G.S., Gakstatter, J.H., and Krawczyk, D.F., Toxicity of sediments from three metal-contaminated areas, *Environ. Toxicol. Chem.*, 3, 279, 1984.

Malueg, K.W., Schuytema, G.S., and Krawczyk, D.F., Effects of sample storage on a copper spiked freshwater sediment, *Environ. Toxicol. Chem.*, 5, 245, 1986.

Markwell, R.D., Connell, D.W., and Gabric, A.J., Bioaccumulation of lipophilic compounds from sediments by oligochaetes, *Water Res.*, 23, 1443, 1989.

Matthews, R.A., Buikema, A.L.J., Cairns, J.J., and Rodgers, J.H.J., Biological monitoring. IIA. Receiving system functional methods, relationships and indices, *Water Res.*, 16, 129, 1982.

Milbrink, G., Biological characterization of sediments by standardized tubificid bioassays, in *Aquatic Oligochaeta*, Brinkhurst, R.O. and Diaz, R.J., Eds., *Hydrobiologia*, 155, 267, 1987.

Monk, D.C., The uses and abuses of ecotoxicology, *Mar. Pollut. Bull.*, 14, 284, 1983.

Mudroch, A. and Azcue, J.M., *Manual of Aquatic Sediment Sampling*, Lewis Publishers, Boca Raton, FL, 1995, 219.

Mudroch, A. and Bourbonniere, R.A., Sediment sample handling and processing, in *Handbook of Techniques for Aquatic Sediments Sampling*, Mudroch, A. and MacKnight, S.D., Eds., CRC Press, Boca Raton, FL, 1994, 131.

Munawar, M. and Munawar, J.F., Phytoplankton bioassays for evaluating toxicity of *in situ* sediment contaminants, *Hydrobiologia*, 149, 87, 1987.

Munawar, M.D., Thomas, R.L., Norwood, W.P., and Daniels, S.A., The toxicity of Detroit River sediment-bound contaminants to ultraplankton, *J. Great Lakes Res.*, 11, 264, 1985.

Munawar, M.D., Gregor, D., Daniels, S.A., and Norwood, W.P., A sensitive screening bioassay technique for the toxicological assessment of small quantities of contaminated bottom or suspended sediments, in *Sediment/Interactions*, Sly, P.G. and Hart, B.T., Eds., *Hydrobiologia*, 176/177, 497, 1989a.

Munawar, M., Munawar, I.F., and Lepard, G.T., Early warning assays: an overview of toxicity testing with phytoplankton in the North American Great Lakes, *Hydrobiologia*, 188/189, 237, 1989b.

Munawar, M., Munawar, I.F., Mayfield, C.I., and McCarthy, L.H., Probing ecosystem health: a multidisciplinary and multi-trophic assay strategy, *Hydrobiologia*, 188/189, 93, 1989c.

Nebeker, A.V., Cairns, M.A., Gakstatter, J.H., Malueg, K.W., Schuytema, G.S., and Krawczyk, D.F., Biological methods for determining toxicity of contaminated freshwater sediments to invertebrates, *Environ. Toxicol. Chem.*, 3, 617, 1984.

Nebeker, A.V., Onjukka, S.T., Cairns, M.A., and Krawczyk, D.F., Survival of *D. magna* and *H. azteca* in cadmium spiked water and sediment, *Environ. Toxicol. Chem.*, 5, 933, 1986.

Nebeker, A.V., Onjukka, S.T., and Cairns, M.A., Chronic effects of contaminated sediment on *Daphnia magna* and *Chironomus tentans*, *Bull. Environ. Contam. Toxicol.*, 41, 574, 1988.

Nebeker, A.V., Schuytema, G.S., Griffis, W.L., Barbitta, J.A., and Carey, L.A., Effect of sediment organic carbon on survival of *Hyalella azteca* exposed to DDT and endrin, *Environ. Toxicol. Chem.*, 8, 705, 1989.

Nelson, M.K., Landrum, P.F., Burton, G.A., Klaine, S.J., Crecelius, E.A., Byl, T.D., Gossiaux, D.C., Tsymbal, V.N., Cleveland, L., Ingersoll, C.G., and Sasson-Brickson, G., Toxicity of contaminated sediment in dilution series with control sediments, *Chemosphere*, 27, 1789, 1993.

Newman, M.C., *Quantitative Methods on Aquatic Ecotoxicology*, Lewis Publishers, Boca Raton, FL, 1995, 426.

Norris, R.H., Rapid biological assessment, selecting reference sites and natural variability, in Proc. Joint South African/Australian Workshop: Classification of Rivers and Environmental Health Indicators, Capetown, 1994, 129.

Organization for Economic Cooperation and Development, *OECD Guidelines for Testing Chemicals, Section 2, Effects on Biotic Systems*, OECD, Paris, 1983.

Othoudt, R.A., Giesy, J.P., Grzyb, K.R., Verbrugge, D.A., Hoke, R.A., Drake, J.B., and Anderson, D., Evaluation of the effects of storage time on the toxicity of sediments, *Chemosphere*, 22, 1991.

Pascoe, D., Brown, A.F., Evans, B.M.J., and McKavanagh, C., Effects and fate of cadmium during toxicity tests with *Chironomus riparius* — the influence of food and artificial sediment, *Arch. Environ. Contam. Toxicol.*, 19, 872, 1990.

Patrick, R., A proposed biological measure of stream conditions, *Theor. Appl. Limnol.*, 11, 299, 1951.

Pellinen, J., Soimasuo, R., Sloof, W., and de Kruijf, H., Toxicity of sediments polluted by the pulp and paper industry to a midge (*Chironomus riparius* Meigen), recent advances in ecotoxicology, *Sci. Total Environ.*, Suppl. 2, 1247, 1993.

Phillips, D.J.H., Bioaccumulation, in *Handbook of Ecotoxicology*, Vol. 1, Calow, P., Ed., Blackwell Scientific, Oxford, 1993, 378.

Phipps, G.L., Ankley, G.T., Benoit, D.A., and Mattson, V.R., Use of the aquatic oligochaete *Lumbriculus variegatus* for assessing the toxicity and bioaccumulation of sediment-associated contaminants, *Environ. Toxicol. Chem.*, 12, 269, 1993.

Phipps, G.L., Mattson, V.R., and Ankley, G.T., Relative sensitivity of three freshwater benthic macroinvertebrates to ten contaminants, *Arch. Environ. Contam. Toxicol.*, 28, 281, 1995.

Power, E.A., Munkittrick, K.R., and Chapman, P.M., An ecological impact assessment framework for decision-making related to sediment quality, in *Aquatic Toxicology and Risk Assessment*, Vol. 14, Mayes, M.A. and Barron, M.G., Eds., American Society for Testing and Materials, Philadelphia, 1991, 48.

Prater, B.L. and Anderson, M.A., A 96-hour sediment bioassay of Duluth and Superior Harbor Basins (Minnesota) using *Hexagenia limbata, Asellus communis, Daphnia magna* and *Pimephales promelas* as test organisms, *Bull. Environ. Contam. Toxicol.*, 18, 159, 1977.

Reichardt, W., Heise, S., and Piker, L., Ecotoxicity testing of heavy metals using methods of sediment microbiology, *Environ. Toxicol. Water Qual.*, 8, 299, 1993.

Reynoldson, T.B., A field test of a sediment bioassay with the oligochaete worm *Tubifex tubifex* (Müller, 1774), *Hydrobiologia*, 278, 223, 1994.

Reynoldson, T.B. and Day, K.E., Freshwater sediments, in *Handbook of Ecotoxicology*, Vol. 1, Calow, P., Ed., Blackwell Scientific, Oxford, 1993, 83.

Reynoldson, T.B., Thompson, S.P., and Bamsey, J.L., A sediment bioassay using the tubificid oligochaete worm *Tubifex tubifex*, *Environ. Toxicol. Chem.*, 10, 1061, 1991.

Reynoldson, T.B., Day, K.E., Clarke, C., and Milan, D., Effect of indigenous animals on chronic end points in freshwater sediment toxicity tests, *Environ. Toxicol. Chem.*, 13, 973, 1994.

Reynoldson, T.B., Bailey, R.C., Day, K.E., and Norris, R.H., Biological guidelines for freshwater sediment based on Benthic Assessment of Sediment (the BEAST) using a multivariate approach for predicting biological state, *Aust. J. Ecol.*, 20, 198, 1995.

Reynoldson, T.B., Rodriguez, P., and Martinez, M., A comparison of reproduction, growth and acute toxicity in two populations of *Tubifex tubifex* (Müller, 1774) from the North American Great Lakes and Northern Spain, *Hydrobiologia*, 334, 199, 1996.

Rosiu, C.J., Giesy, J.P., and Kreis, R.G., Jr., Toxicity of vertical sediments in the Trenton Channel, Detroit River, Michigan, to *Chironomus tentans* (Insecta: Chironomidae), *J. Great Lakes Res.*, 15, 570, 1989.

Ross, P.E. and Henebry, M.S., Use of four microbial tests to assess the ecotoxicological hazard of contaminated sediments, *Toxicol. Assess.*, 4, 1, 1989.

Ross, P.E., Jarry, V., and Sloterdijk, H., A rapid bioassay using green alga *Selenastrum capricornutum* to screen for toxicity in St. Lawrence River sediment elutriates, in *Functional Testing of Aquatic Biota for Estimating Hazards of Chemicals*, Cairns, J., Jr. and Pratt, J.R., Eds., ASTM STP 988, American Society for Testing and Materials, Philadelphia, 1988, 68.

Samoiloff, M.R., Bell, J., Birkholz, D.A., Webster, G.R.B., Arott, E.G., Pulak, R., and Madrid, A., Combined bioassay–chemical fractionation scheme for the determination and ranking of toxic chemicals in sediments, *Environ. Sci. Technol.*, 17, 329, 1983.

Santiago, S., Thomas, R.L., Larbaigt, G., Rossel, D., Echeverria, M.A., Tarradellas, J., Loizeau, J.L., McCarthy, L., Mayfield, C.I., and Corvi, C., Comparative ecotoxicity of suspended sediment in the lower Rhone River using algal fractionation, Microtox and *Daphnia magna* bioassays, *Hydrobiologia*, 252, 231, 1993.

Sasson-Brickson, G. and Burton, G.A., *In situ* and laboratory sediment toxicity testing with *Ceriodaphnia dubia*, *Environ. Toxicol. Chem.*, 10, 201, 1991.

Schmidt-Dallmer, M.J., Atchinson, G.J., Steingraeber, M.T., and Knights, B.C., A sediment suspension system for bioassay with small aquatic organisms, *Hydrobiologia*, 245, 157, 1992.

Schubauer-Berigan, M.K. and Ankley, G.T., The contribution of ammonia, metals and nonpolar organic compounds to the toxicity of sediment interstitial water from an Illinois River tributary, *Environ. Toxicol. Chem.*, 10, 925, 1991.

Schubauer-Berigan, M.K., Amato, J.R., Ankley, G.T., Baker, S.E., Burkhard, L.P., Dierkes, J.R., Jenson, J.J., Lukasewycz, M.T., and Norberg-King, T.J., The behavior and identification of toxic metals in complex mixtures: examples from effluent and sediment pore water toxicity identification evaluations, *Arch. Environ. Contam. Toxicol.*, 24, 298, 1993.

Schuytema, G.S., Nebeker, A.V., Griffis, W.L., and Miller, C.E., Effects of freezing on toxicity of sediments contaminated with DDT and endrin, *Environ. Toxicol. Chem.*, 8, 883, 1989.

Smith, D.P., Kennedy, J.H., and Dickson, K.L., An evaluation of a naidid oligochaete as a toxicity test organism, *Environ. Toxicol. Chem.*, 10, 1459, 1991.

Southerland, M.T. and Stribling, J.B., Status of biological criteria development and implementation, in *Biological Assessment and Criteria: Tools for Water Resource Planning and Decision Making;* Davis, W.S. and Simon, T.P., Eds., Lewis Publishers, Boca Raton, FL, 1995, 81.

Stemmer, B.L., Burton, G.A., Jr., and Leibfritz-Frederick, S., Effect of sediment test variables on selenium toxicity to *Daphnia magna*, *Environ. Toxicol. Chem.*, 9, 381, 1990a.

Stemmer, B.L., Burton, G.A., Jr., and Sasson-Brickson, G., Effect of sediment spatial variance and collection method on cladoceran toxicity and indigenous microbial activity determinations, *Environ. Toxicol. Chem.*, 9, 1035, 1990b.

Stephenson, G.L., Day, K.E., Scroggis, R., Kirby, R.S., and Hamr, P., Current status of Environment Canada's guidance on control of test precision using spiked sediment toxicity test: what to spike and how? in *Environmental Toxicology Risk Assessment*, Vol. 4, La Point, T.W. and Ingersoll, C.G., Eds., American Society for Testing and Materials, Philadelphia, 1995.

Suedel, R.C. and Rodgers, J.H., Development of formulated reference sediment for freshwater and estuarine sediment testing, *Environ. Toxicol. Chem.*, 13, 1163, 1994.

Suedel, B.C., Rodgers, J.H., and Clifford, P.A., Bioavailability of fluoranthene in freshwater sediment toxicity tests, *Environ. Toxicol. Chem.*, 12 , 155, 1993.

Taub, F.B., Standardized aquatic microcosms, *Environ. Sci. Technol.*, 23, 1064, 1989.

Taylor, E.J., Rees, E.M., and Pascoe, D., Mortality and a drift-related response to the freshwater amphipod *Gammarus pulex* (L) exposed to natural sediments, acidification and copper, *Aquat. Toxicol.*, 29, 83, 1994.

Touart, L.W., Hazard Evaluation Division, Technical Guidance Document: Aquatic Mesocosm Tests to Support Pesticides Registration, Office of Pesticide Programs, U.S. Environmental Protection Agency, Washington, D.C., 1988.

U.S. Environmental Protection Agency, Short-Term Methods for Estimating the Chronic Toxicity of Effluents and Receiving Waters to Freshwater Organisms, 2nd ed., U.S. EPA, Cincinnati, 1989.

U.S. Environmental Protection Agency, Methods for Measuring the Acute Toxicity of Effluents and Receiving Waters to Freshwaters and Marine Organisms, 4th ed., U.S. EPA, Cincinnati, 1991.

Van Hattum, B., Korthals, G., VanStraalen, N.M., Govers, H.A., and Joosse, E.N.G., Accumulation patterns of trace metals in freshwater isopods in sediment bioassays: influence of substrate characteristic, *Water Res.*, 27, 669, 1993.

Warren, C.E., *Biology and Water Pollution Control*, W.B. Saunders, Philadelphia, 1971, 434.

West, C.W., Mattso, V.R., Leonard, E.N., Phipps, G.L., and Ankley, G.T., Comparison of the relative sensitivity of 3 benthic invertebrates to copper: contaminated sediments from the Keweenaw Waterway, *Hydrobiologia*, 262, 57, 1993.

West, C.W., Phipps, G.L., Hoke, R.A., Goldenstein, T.A., Vandermeiden, F.M., Kosian, P.A., and Ankley, G.T., Sediment core versus grab samples: evaluation of contamination and toxicity at a DDT contaminated site, *Ecotoxicol. Environ. Saf.*, 28, 208, 1994.

Whitton, B.A., *River Ecology*, Blackwell Scientific, Oxford, 1975, 725.

Wiederholm, T. and Dave, G., Toxicity of metal polluted sediments to *Daphnia magna* and *Tubifex tubifex*, *Hydrobiologia*, 176/177, 411, 1989.

Wiederholm, T., Wiederholm, A.M., and Milbrink, G., Bulk sediment bioassays with five species of fresh-water oligochaetes, *Water Air Soil Pollut.*, 36, 131, 1987.

Winger, P.V. and Lasier, P.J., Sediment toxicity in Savannah Harbor, *Arch. Environ. Contam. Toxicol.*, 28, 357, 1995.

Wong, P.T.S., Chau, Y.K., Ali, N., and Whittle, D.M., Biochemical and genotoxic effects in the vicinity of a pulp mill discharge, *Environ. Toxicol. Water Qual.*, 9, 59, 1994.

Wright, J.F., Moss, D., Armitage, P.D., and Furse, M.T., A preliminary classification of running water sites in Great Britain based on macro-invertebrate species and the prediction of community type using environmental data, *Freshwater Biol.*, 14, 221, 1984.

Yoder, C.O. and Rankin, E.T., Biological criteria program development and implementation in Ohio, in *Biological Assessment and Criteria: Tools for Water Resource Planning and Decision Making*, Davis, W.S. and Simon, T.P., Eds., Lewis Publishers, Boca Raton, FL, 1995, 109.

Zagatto, P.A., Gherardi-Goldstein, E., Bertoletti, E., Lombardi, C.C., Martins, M.H.R.B., and Ramos, M.L.L.C., Bioassays with aquatic organisms: toxicity of water and sediment from Cubatao River Basin, *Water Sci. Technol.*, 19, 95, 1987.

chapter four

Field methods and interpretation for sediment bioassessment

Trefor B. Reynoldson and Pilar Rodriguez

4.1 Introduction

The previous chapter of this book provided the rationale for incorporating biological methods in sediment assessments and presented an overview of laboratory methods for assessing sediment toxicity. In this chapter, the authors describe the various field methods available for assessing sediment quality. There are three major categories of methods available for conducting assessments (Chapman et al., 1992):

- *Tissue chemistry analysis*: Identifies the presence and concentrations of measured chemicals in organisms and tissues and provides an indication of bioavailability. However, it does not provide information on effects, the presence of biologically transformed chemicals, or the presence of chemicals which were not measured.
- *Pathological studies*: Identify the presence and degree of response in populations, organisms, or tissues. However, they do not identify the effect of the pathological condition at a higher level of organization or unmeasured responses in populations, organisms, or tissues.
- *Community structure studies*: Identify the presence and abundance of populations, taxa, and individuals. However, they do not provide information on causality.

Bioassessment by nature relates to a specific and particular set of circumstances and, unlike anticipatory testing of new chemicals, is not practicable to provide a standardized approach or set of methods. This is an important and often misunderstood difference, particularly by regulators who are used to standard testing protocols between the two types of testing

1-56670-343-3/99/$0.00+$.50

necessary to establish the biological effects of chemicals. Therefore, one cannot recommend a suite of four or five attributes to be measured. Rather, the study designer should be aware of the array of tools available and their advantages and shortcomings and select appropriate measures accordingly.

In an attempt to assess the relative usefulness of the methods and provide advice on their general utility, the authors have used five criteria adapted from Giesy and Hoke (1989) and Calow (1993a):

- *Relevance*: The ecological significance of the measurement at scales greater than that being tested. In all testing, there is concern about the overall biological structure and function. The authors could not test all system attributes, and therefore the parameter being tested is simply a surrogate for the system as a whole. The ease with which one can relate the test to the system is a measure of *relevance*. In many cases, this cannot be done quantitatively but rather is a subjective evaluation.
- *Robustness*: To provide consistency in interpretation and decision making, it is important that the tests provide consistent results irrespective of when and where they are conducted. This is clearly more difficult in the case of field than laboratory tests, where environmental variation can be large. However, standardized methods and protocols can reduce a significant portion of the variation (Figure 4-1, example of pulp mill data). Some measurement variables are inherently more robust in either or both time and space.
- *Methodology*: Both ease of use and detectability of the measurement end point (or availability of test organisms in laboratory tests) are important criteria. If the method is too complex, expensive, or of an inappropriate duration, it is of limited value for routine use.
- *Sensitivity*: The variable chosen for measurement should be sensitive to toxicants and in an appropriate range, sufficient to detect a response but also to discriminate between measurement locations.
- *Appropriateness/application*: The method being selected should be appropriate to the system for which the study is being designed. Many useful methods are only applicable to specific conditions or environments. Some methods have a much broader application than others.

In the real world, no single test method maximizes all these attributes, and there must be trade-offs between the selection criteria. However, at all times each criterion should be considered when selecting methods, and a number of tests are of consequence necessary. Field methods described in this chapter tend to be more powerful regarding relevance and application and less powerful in the sense of robustness, which is the strength of laboratory testing. This is the reason the authors and other researchers strongly argue for both a field- and laboratory-based approach.

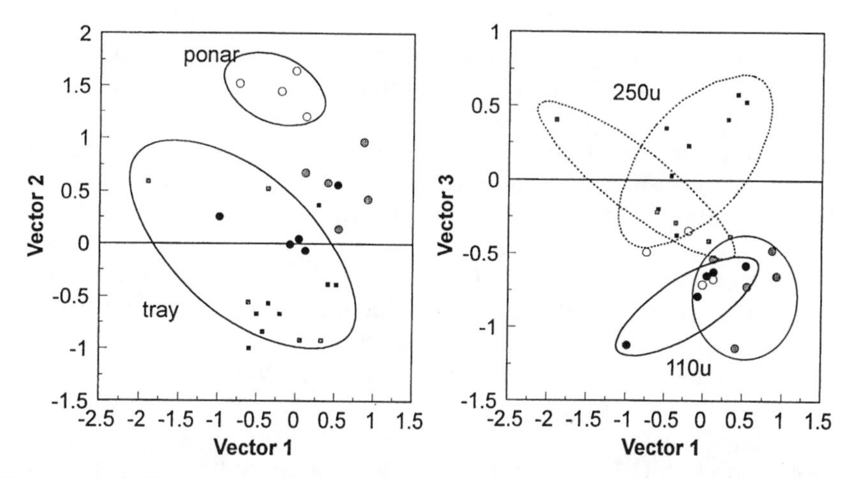

Figure 4-1 Effect of sampling device and sample processing on interpretation of benthic invertebrate community structure data up- and downstream of a pulp mill effluent: 1975 — Ponar 250-µm mesh (hatched squares), 1977 — tray 250-µm mesh (■), 1984— tray 110-µm mesh (●), 1985 — tray 110-µm mesh (hatched circles).

This chapter describes the various methods available. It also describes a framework for the selection of appropriate methods, a guide to source material, and provides an introduction to methods available for data interpretation.

4.2 Overview and assessment of field test methods

4.2.1 Tissue analysis

4.2.1.1 Direct measurement

In this case, we are addressing only the issue of tissue analysis in field-collected animals for the presence of contaminants. This method provides information on exposure of the animal to contaminants. However, it does not indicate that there is an effect. There are two basic approaches to assessing field exposures: (1) the collection and measurement of field-collected organisms and (2) exposing collected animals to contaminants for a fixed time period.

In adopting the "sentinel organism" approach, both Phillips (1980) and Hellawell (1986) identified the following attributes that an ideal sentinel organism should have:

1. Individuals of a species should show the same correlation between their contaminant content and the average concentration of the contaminant in the sediment at all locations and under all conditions.

2. Individuals of a species should not be killed or rendered incapable of long-term reproduction by maximum levels of the contaminant.
3. The species should be sedentary.
4. The species should be large enough or present in sufficient numbers to provide enough tissue for analysis.
5. Individuals of a species should be widespread enough to facilitate comparisons among areas.
6. The species should be sufficiently long-lived to enable sampling of several year classes to provide evidence of long-term effects.
7. Individuals should be easy to collect.
8. Individuals should be hardy enough to survive laboratory or field handling (or caging).

Johnson et al. (1993) suggested that there are two basic approaches to the use of sentinel organisms in survey and experimental studies. The former provides information on the current status of contaminants in the ecosystem over a wide geographic area. The latter tends to be site specific, often with repeated sampling to provide time–trend data. In all these types of studies, it is usually necessary to calibrate the organisms as the first requirement (described in item 1 above) is rarely met. Calibration involves establishing the effects of abiotic and biotic variables (Table 4-1) on the uptake of contaminants and relates the concentration in the organism to the environment (Jones and Walker, 1979; Millington and Walker, 1983; Russell and Gobas, 1989; Muncaster et al., 1990).

4.2.1.2 Bioaccumulation models

The simplest method of measuring exposure is measurement of contaminant residue in the tissues of field or laboratory-exposed animals. However, these approaches for monitoring bioaccumulation by sediment-dwelling organisms are costly and have limited ability to predict tissue residues that

Table 4-1 Confounding Factors for Assessing Bioaccumulation Through Tissue Analysis

Abiotic factors	Biotic factors
Temperature	Lipid content
Water chemistry	Uptake rate
Sediment geochemistry	Depuration rate
Contaminant form	Tolerance
	Age
	Size and weight
	Growth rate
	Diet
	Sex
	Reproductive condition

result from changes to the exposure level (Lee, 1992). An alternate approach is the use of sediment bioaccumulation models, of which there are two general types: equilibrium-based and kinetic models. The former assume steady-state conditions between the organisms and the environment. Kinetic approaches describe tissue concentration as the net effect of rate processes. However, as this approach is not a field method, the authors have not attempted to review the various methods.

4.2.2 *Pathology*

In the context of sediment bioassessment, the authors define pathological studies as those methods that identify the presence or level of a functional response in processes, tissues, organisms, populations, or communities. These types of tests cover the entire range of biological organization, from the simplest (e.g., the cell) to most complex (e.g., ecosystem). As Calow (1993b) states, "Moving down the hierarchy (ecosystem to molecule), systems become more easy to control, reaction times reduce and criteria for robustness, methodology and sensitivity improve, and generality should also increase. However, since ecological principles largely involve individuals and groups of individuals relevance (and application) increase in the opposite direction." In selecting tests for application, the operator has to balance these two dilemmas and inevitably compromise decisions have to be made. In the following pages, the authors provide an overview of the wide array of tests available. Four broad test categories are distinguished along an organizational gradient in Table 4-2, which provides examples of end points that can be measured and identifies comprehensive reviews of the test categories which can be used to access the relevant literature.

4.2.2.1 *Cellular bioindicators*

The advantage of cellular bioindicators (biomarkers) is that their response shows that contaminants are present in the sediment, are available to the organism, and have reached the measured tissue in a sufficient quantity and for a sufficient duration to produce a response. The merit of the measured responses is progressive, from a simple homeostatic response through initiation of tissue damage to organ damage, death of the individual, and, finally, effects on the population. Because many of these tests are highly specific, they can have value as contaminant-specific responses (e.g., content of metallothionein as an indicator of metals), but consequently they require a battery of tests to be used to examine different contaminant classes.

A number of cellular bioindicators of contaminant exposure are presently being used in environmental monitoring. Some of these have been extensively field validated and are in routine application:

- Inhibition of brain or blood cholinesterase for pesticides
- Induction of hepatic microsomal cytochrome P450 for polynuclear aromatic hydrocarbons (PAHs) and polychlorinated biphenyls (PCBs)

Table 4-2 Summary of Field Methods for Assessing Pathology

Method category	Example end points	Reviews
Cellular bioindicators	Cholinesterase Metallothionein Heat stress proteins	McCarthy and Shugart, 1990; Benson and Di Giulio, 1992; Hugget et al., 1992; Stegeman et al., 1992; Weeks et al., 1992; Peakall, 1992; Melancon, 1995
Organism bioindicators	Growth Reproduction Behavior Histopathology	Baudo et al., 1990; Mac and Schmitt, 1992; Johnson et al., 1993; La Point, 1995
Population effects	Life histories Genetic variability	Diaz, 1992; La Point and Fairchild, 1992; Reice and Wohlenberg, 1993
Ecosystem indicators	Trophic levels Energy flow Nutrient cycling Succession	Burgess and Scott, 1992; Cairns et al., 1992; Diaz, 1992

- Reproductive aberrations such as terata and eggshell thinning (these tests have been primarily used in avian and other vertebrate species and thus are of limited value for sediment bioassessment)
- Aberrations in hemoglobin synthesis (porphyrin) for lead and hydrocarbons

There has been considerable work on tests using DNA damage and histopathological effects, particularly tumors, as end points. The measurement of metallothionein and stress proteins requires further development of more convenient and inexpensive methodologies, as well as field validation, but holds good promise. The various available cellular biomarkers are elaborated in Table 4-3, which also indicates the degree of specificity of each test, appropriate literature for each test, and an evaluation of the test for the criteria the authors consider important in selection of the test.

Much of the research on cellular-level indicators has been conducted on terrestrial species, particularly birds and mammals. The major aquatic group examined has been fish. Furthermore, many of the methods have not been validated in the field, and the effects of normal environmental stresses, such as temperature and food supply, have not been established. This is particularly the case for some of the biochemical (e.g., metallothionein, stress proteins, and oxyradicals) and physiological (e.g., corticosteroids, glycogen, lipids, and RNA:DNA) indicators.

The authors would suggest, at the present time, that the most immediately useful tests are acetylcholinesterase, P450, and DNA damage. Tests that show promise and which should be investigated are porphyrins, metallothionein, and heat-stress proteins.

4.2.2.2 Organism bioindicators

At the organism level, the authors have identified three major test types (Table 4-4): histopathological changes, which primarily involve the assessment of neoplasms in organisms; physiological measurements; and the occurrence of morphological deformities.

The most widely used tests on field-collected animals are histopathological measures. Biochemical or physiological alterations, if severe enough or existing for a long enough duration, can result in structural alterations in organelles, cells, tissues, and organs. The detection of alterations to such structures may be indicative of exposure to chemical contaminants. The existence of neoplasms was one of the first uses of histopathological indices. The primary tissue of investigation has been the liver. Liver tissue neoplasms in fish (Dawe et al., 1964) have been directly linked to sediment contamination (Pierce et al., 1978). Similar links between neoplasms in epidermal tissue in fish and contaminated sediments have been made (Black, 1983). Bottom-feeding fish are the most likely candidate species for investigation of neoplasms as indicators of sediment contamination.

Organism-level physiological tests rely on measures of change in either respiration, growth, or reproduction. Most of these studies are conducted in the laboratory with field-collected individuals and therefore are not included in this chapter. However, they are described in the previous chapter of this book. Similarly, behavioral and ecological measurements at the organism level are primarily laboratory-based tests at the organism level and are also described in the preceding chapter.

Morphological deformities in response to contaminants have been identified in an array of vertebrate and invertebrate species. For example, the most commonly seen deformities in birds are modifications to claws and beaks. In fish, spinal deformities have been linked to the presence of contaminants. However, for assessing specifically sediment-associated contaminants, benthic invertebrate species are the best indicator organisms. The occurrence of deformities in benthic invertebrates has been known for more than 20 years (Brinkhurst et al., 1968). Two groups of animals have been shown to demonstrate morphological anomalies: the oligochaete and the chironomidae. Warwick (1988) has extensively reviewed the use of deformities in chironomids and observed that the frequency of deformity occurrence varies widely. Present methods only provide a qualitative measurement of contaminant effects. Johnson et al. (1993) concluded the following: (1) the frequency of occurrence in response to contaminant gradients must be established, (2) the full range of deformities needs documenting, (3) the

Table 4-3 Available Cellular Bioindicators for Application in Sediment Bioassessment

Method	Specificity	Relevance	Robustness	Methodology	Sensitivity	Application	Reference
Biochemical							
Cholinesterase	Pesticides	1	1	2	2	1	Ludke et al., 1958; Weiss, 1958, 1961; Gibson et al., 1969; Fairbrother et al., 1991; Rattner and Fairbrother, 1991; Edwards and Fisher, 1991; Peakall, 1992; Mayer et al., 1992
Cytochrome P450	?	1	1	2	2	1	Burke and Mayer, 1974; Van Contfort et al., 1977; Klotz et al., 1984; Burke et al., 1985; Stegeman et al., 1987; Goksoyr, 1991
Metallothionein	Specific metals	0	0	1	1	1	Hamer, 1986; Engel and Brouwer, 1989; Benson et al., 1990; Petering et al., 1990; Dutton et al., 1993
Stress proteins	Broad	0	0	1	0	0	Sanders, 1990, 1993
Oxyradicals	Broad	0	0	1	0	0	Di Giulio et al., 1989; Di Giulio, 1991; Benson and Di Giulio, 1992; Stegeman et al., 1992
Genotoxic							
DNA damage	PAHs, aromatic amines, carbazoles	1	0	2	2	1	Randerath et al., 1981; Dunn et al., 1987; Shugart, 1988, 1990; Varanasi et al., 1989; Bickham, 1990; McMahon et al., 1990; Peakall, 1992; Shugart et al., 1992; Meyers-Schone et al., 1993
Immunological							
Phagocytosis	?	0	0	1	0	0	Warriner et al., 1988; Weeks et al., 1992
Lysozome activity	?	0	0	1	0	0	Sharma, 1981; Weeks et al., 1992; Wong et al., 1992

Physiological							
Hemoglobin ALAD and porphyrins	Lead	1	1	2	0	0	Galen, 1975; Lockhart and Metner, 1984; Kennedy et al., 1986; Reddy et al., 1987; Watson et al., 1989; Melancon et al., 1992; Peakall, 1992
Corticosteroids	?	0	0	2	1	0	Donaldson and Dye, 1975; DiMichelle and Taylor, 1978; Bry, 1982; Thomas, 1982
Glycogen	?	1	0	2	1	0	Bayne, 1973a,b; Murty and Devi, 1982; Gill and Pant, 1983; Verma et al., 1983; Chaudry, 1984
Lipids	?	1	0	2	0	0	Gardner and Riley, 1972; Fletcher, 1984; Dey et al., 1983; Dillon and Benson, 1987
RNA:DNA	?	0	0	1	0	0	Sutcliffe, 1970; Bulow, 1987; Kearns and Atchison, 1979

Table 4-4 Methods for Field Pathological Assessment at the Level of the Organism, Population, and Ecosystem

	Relevance	Robustness	Methodology	Sensitivity	Application	Reference
ORGANISM						
Histopathological						
Neoplasms	1	1	2	0	1	Black, 1983; Fabacher and Baumann, 1985; Hinton et al., 1992
Physiological						
Respiration	1	0	1	0	1	Maki et al., 1973; Wiederholm et al., 1987; Coler
Growth	2	0	1	0	2	et al., 1988; Naylor et al., 1989
Morphological						
Setae	0	0	1	0	0	Milbrink, 1983
Mouthparts	0	0	1	0	0	Warwick, 1985, 1991
POPULATION						
Behavioral and ecological						
Swimming activity	1	0	1	0	0	Detra and Collins, 1991
Net spinning	1	0	1	0	0	Petersen and Petersen, 1984
Burrowing	1	0	1	0	0	Heinis et al., 1990
Reproduction	2	1	1	1	2	Henry et al., 1986
Feeding	2	1	1	1	2	Keilty and Landrum, 1990
Sediment processing	0	0	1	0	2	Hadderingh et al., 1987
Recruitment mode and time	2	1	1	0	0	
Habitat preference	1	0	1	0	0	
ECOSYSTEM						
Carbon uptake	2	0	0	0	2	Sayler et al., 1979; Jones and Hood, 1980; Medine
O_2 production/demand	2	0	0	0	2	et al., 1980; Gahnstrom, 1985; Burton et al., 1987;
Decomposition	2	0	0	0	2	Munawar and Munawar, 1987; Ross et al., 1988;
Nutrient cycles	2	0	0	0	2	Scanferlato and Cairns, 1990

methods need simplifying (i.e., indices proposed by Warwick [1985, 1991] are highly technical), and (4) cause-and-effect relationships need to be established. At the present time, the authors cannot recommend the use of deformities as a standard tool, primarily because of their specialized nature.

4.2.2.3 Population level

Effects that can be measured at the population level are associated with survival and reproduction of the individual. While these can be measured using individuals in laboratory-based tests (see above), they can also be estimated from counts of individuals collected in the field. Population age and attributes, such as the intrinsic rate of increase, can be calculated and used as an indicator of stress (Pasteris et al., 1996). However, these functional attributes of communities have not been widely used for assessment purposes, and most population-level field methods are structural in nature. They are described in the following section on community structure assessment.

The genetic variation of a population may also be indicative of environmental stress. The development of a physiological or behavioral response to pollution gradients may well result in inherited resistance in a population. There is evidence for genetically distinct populations of *Chironomus* in response to oxygen stress (Pedersen, 1984) and in the oligochaete *Limnodrilus hoffmeisteri* in response to cadmium and nickel exposure (Klerks and Levinton, 1989). If the genetic range of reference populations could be established, then this could provide a baseline for estimating response in stressed populations.

Ecological or genetic methods at the population level have not been used widely in impact assessment. Also, they do not focus on sediment as the medium of impact. At this time, the authors do not believe that these methods should be considered for routine use.

4.2.2.4 Ecosystem level

Ecosystem-level testing of functional attributes is simulated either by conducting *in situ* assays or by measuring functional end points that integrate ecosystem-level processes. *In situ* assays are a relatively new approach to assessing sediment contamination. While this approach was proposed more than 10 years ago (Nebeker et al., 1984), the authors are unaware of its widespread adoption. This is primarily associated with the lack of control, cost, and usual problems of retrieval success associated with field-deployed equipment. Studies that have been conducted have shown significant differences between *in situ* and laboratory responses (Sasson-Brickson and Burton, 1991). Furthermore, these *in situ* tests are largely conducted with a single species and, therefore, are only considered to be at the ecosystem level because they are conducted in the field.

A number of ecosystem-level functions can be measured to provide an estimate of the effects of anthropogenic stress. Measures of ecosystem function are usually estimates of the rate of energy flow and material cycling or biological regulation of the environment using end points such as respiration, mineralization, and nutrient regeneration (Cairns et al., 1992). While there is clearly an intimate association between structural end points (i.e., kinds and numbers of organisms) and functional end points, which are the most sensitive functional tests is not evident. There is little empirical evidence available on the relationship between structural and functional measurements. For the purposes of sediment assessment, most of the existing work has been focused on microbial activity. Results of studies showed effects of contaminated sediments on primary production (Ross et al., 1988; Munawar and Munawar, 1987), oxygen production and respiration (Medine et al., 1980), and decomposition. Similarly, effects of cycling of both nitrogen (Prichard and Bourquin, 1985) and phosphorus (Burton and Stemmer, 1988) were demonstrated. Despite the potential importance of the detection of functional impairments, these methods are not a common component of field studies, and additional background information is necessary to make these end points useful additions to the structural measures used in routine assessments of sediment contamination.

4.2.3 *Community structure*

While a number of different groups of organisms, such as bacteria, algae, and fish, are associated with bottom sediments, it can be argued that the most appropriate group for investigating sediment contamination is benthic invertebrates, and they are the most widely used (Rosenberg and Resh, 1993; Resh et al., 1995; Southerland and Stribling, 1995). This assemblage of organisms covers a broad spectrum of trophic groups, exhibits a variety of strategies for feeding and shelter that result in intimate association with bottom sediments, and therefore most likely exhibits effects associated with sediment contamination. Other attributes of this group which make it ideal for sediment assessment are (1) they are spatially representative, (2) their life cycles are of an appropriate temporal scale so that repeated sampling is not necessary, (3) their taxonomy is well understood, and (4) there is extensive information in the scientific literature on their use in biological monitoring. Consequently, this section of the chapter deals exclusively with issues pertaining to the use of benthic invertebrates in sediment assessments.

4.2.3.1 *Study design considerations*

The remainder of this section deals with the issues that should at least be considered prior to and after conducting a community structure survey. Those involved in designing and executing the field collection of invertebrates should at least be aware of these issues. The authors have not attempted to be inclusive, as books have been written on this one issue. They have, however, attempted to raise the consciousness on each of the important steps.

There are a number of steps in designing a sampling program for sediment bioassessment (Norris and Georges, 1993). Each of these is considered separately as listed below:

Sample Collection and Field Measurements
↓
Sample Preservation and Transport
↓
Sample Processing and Analysis
↓
Data Recording, Transcription, and Entry
↓
Data Analysis and Interpretation

4.2.3.2 Sample collection and field measurements

In some regards, this is the most important stage of a valuable assessment. Once this step is completed, there is no going back and errors made cannot be corrected. Resh and McElravy (1993) identified the following decisions that must be made for the sample collection and field measurement phase.

4.2.3.2.1 Number of replicates. This is probably the most contentious and widely discussed area in study design. The collection of replicates is carried out for two reasons. The first is to estimate the value of a measurement with a desired and known degree of precision. The second is to determine if a given degree of change in a measure has occurred among several sites (or times) with a known risk of error. A number of factors will influence the number of replicates to be collected for estimating the mean of a benthic measurement: the size of the mean value, the degree of aggregation, and the desired precision. A detailed discussion and methods for identifying the number of replicates can be found in Needham and Usinger (1956), Chutter (1972), Elliott (1977), Green (1979), and Resh (1979). Low values of the mean, greater degrees of aggregation, and higher desired precision each require greater numbers of sample units. Pilot studies are essential for establishing that the level of precision is appropriate to the study objectives.

4.2.3.2.2 Stratification To maximize the use of resources in a study design, the authors would strongly recommend stratifying the habitat to be sampled. This can be carried out on habitat characteristics, such as grain size or depth, or with regard to treatment and control. However, a choice of some level of sample stratification is important for reducing intersample variability.

4.2.3.2.3 Sampling device. A plethora of sampling devices are available for sampling soft sediments. One of the most comprehensive reviews of available sampling devices was conducted by Elliott and Tullett (1978), in which they identified sampling devices according to their operation (i.e., shallow or deep), the substrate, and whether or not they were developed for freshwater or marine use. Reference to samplers for fine-grained sediment habitats is the largest category they identify, with a total of 329 references listed. Several basic types of sampling devices are listed: nets and quadrat samplers (5); scoops, shovels, and dredges (56); grabs (63); corers (159); suction and air lifts (41); and electroshocking (5). The largest single group of references describes small-diameter corers (i.e., <10 cm). In a review of lentic studies (Resh and McElravy, 1993), the three most frequently used devices were grabs and dredges (i.e., 52%), corers (i.e., 26%), and scoops (i.e., 10%). It is difficult to recommend a specific sampling device. The choice depends on the objectives of the study. However, considerations should be given to cost, availability, and other logistical factors. Sampler accuracy and precision should be established in a pilot study, and ideally several comparable devices should be considered. The maxim more small replicates is best should be borne in mind. For example, seven samples collected by benthos corer, each 6.6 cm in diameter, can be processed with the same effort as one sample collected by a 15-cm Ekman grab sampler.

4.2.3.2.4 Mesh size. In 45 lentic studies, a mesh size of 101 to 200 μm was most commonly used (i.e., in 23% of the studies). Only 15 of the studies used a mesh size greater than 500 μm (Resh and McElravy, 1993). This is considerably smaller than the 450- to 600-μm range of the mesh size reported by Downing (1984) as being most common in lentic studies. Again, the objectives of the study are important in the selection of the mesh size. If the study is focused on population dynamics or life history, where capture of small, young animals is critical, then small mesh sizes (i.e., 250 μm or smaller) are recommended. If the objective is to characterize a site taxonomically, then larger mesh sizes (e.g., 400 to 600 μm) are appropriate. The time saved, and thus the number of samples that can be processed using larger mesh sizes, is considerable.

4.2.3.3 Sample preservation and transport

Once the sediment samples are recovered, the authors recommend sieving in the field, if appropriate conditions are available, as live material is less fragile. However, if this is not possible, samples should be preserved in buffered formalin for a few days and then transferred to ethyl alcohol. After sieving, benthos samples should be preserved in alcohol prior to final counting and identification. Ideally, the same method should be used throughout the study. The possibility that preserving the sample prior to sieving affects the flexibility of the organisms and how they are retained by sieves should be verified. The authors strongly recommend the use of buffered formalin, as the initial use of alcohol results in poorly preserved oligochaetes. This group is often an important component of the community in fine-grained sediment environments. Furthermore, buffered formalin is recommended; otherwise, the calcium in mollusc shells can be leached out, and the shells become extremely fragile. Appropriate sample labeling is paramount. Sample containers should be labeled on the outside of the container and on the container lid, and a paper label should be placed inside the container. The label should indicate the sample location, the replicate number, and the sampling date as a minimum.

4.2.3.4 Sample processing and analysis

4.2.3.4.1 Subsampling. A number of methods have been used to reduce the time required to process samples. These include processing only selected samples, combining samples, or subsampling. In two separate surveys, the authors (Resh et al., 1985; Resh and McElravy, 1993) obtained quite different results, indicating that different groups process samples differently. Academics and research biologists tend to process all sample units, whereas consulting and industry biologists tend to process only a part of the samples. Subsampling can legitimately be used if the error associated with the subsampling method has been established beforehand. A number of subsampling methods are available (Hilsenhoff, 1977, 1982; Courtemanch, 1989; Marchant, 1989; Plafkin et al., 1989). These are usually based on selecting a predetermined number of organisms (e.g., 100, 200, 300) or a predetermined time (e.g., 30 min). The critical factor in using such a method is removing bias from the selection of the portion of the sample to be examined. This is normally carried out using a marked tray or a subdivided box (Marchant, 1989) and randomly selecting cells until the subsampling criterion (i.e., number or time) is met. Other methods used simply involve dividing the sample into a half or a quarter. However, such methods have not been verified. Whichever method is adopted should be tested against several whole samples to verify the adequacy of the method in estimating the whole sample before adopting it as a standard procedure (e.g., see Sebastien et al., 1988; Wrona et al., 1982).

4.2.3.4.2 Picking. A number of procedures can be used to facilitate the removal of organisms from the sample. These methods are more typically required with lotic samples, which tend to be more heterogenous and have more debris than fine-grained sediment samples. However, if required, available methods include elutriation, flotation, and sieving. Resh et al. (1985) have shown the time savings associated with the methods to be 34.7% for elutriation or flotation and 24.8% for staining. These are significant savings, and the use of such methods is worth consideration in samples where separating the animals is time consuming.

4.2.3.4.3 Identification. The prevailing view has been that species or lowest possible taxonomic level identifications are critical in biomonitoring studies (Resh and Unzicker, 1975). However, there is little information on the critical assessment of the level of taxonomic detail required to detect disturbance in freshwater ecosystems. There has been some assessment of the level of taxonomic discrimination required in marine systems. Warwick (1988) reviewed five data sets at the species, family, and phylum levels in response to pollution gradients using both multivariate and univariate analyses. In no case was there any substantial loss of information at the family level. Furthermore, site groupings in the multivariate analyses for macrofauna data (i.e., >500 μm) related more closely to the pollution gradient at the phylum level than the species level in two of three studies and was no worse than the species level in the third one. Warwick (1988) suggested that the reason higher level taxonomic discrimination is suitable for identifying anthropogenic effects is that such changes modify community composition at a higher level than environmental variables, which alter communities by species replacement. No similar study has been undertaken for freshwater environments, and it is likely that taxonomic levels lower than phylum are required, given the much lower number of phyla in freshwater communities. However, a thorough examination of the taxonomic level required may well show that the species-level identifications are not required and that the family level may be sufficient to detect pollution gradients.

4.2.3.5 Data recording, transcription, and entry

Errors that have occurred in data through sample collection, field measurement, preservation, and sample transport are usually irretrievable, and procedures should be put in place to minimize them. However, errors can also occur in data recording, transcription, and entry, and procedures should also be adopted to minimize them. Because assessment studies frequently involve the production of large databases, some methods are required for screening data. The following procedures, suggested by Norris and Georges (1993), form a good basis for ensuring that preventable errors do not occur:

- Examination of the ranges in numbers of individuals and species per replicate for exceedences
- Calculation of means and standard deviations for subsets (e.g., by date or site) for aberrantly high or low values
- Univariate checks for outliers, by testing for values more than three standard deviations outside the mean
- Plots of variables against each other or against time or distance
- Checks on total counts can verify that counts of individual species have been entered correctly

It is almost inevitable that such exploratory analysis will reveal errors and missing data. At this point, well-designed field and laboratory data entry sheets will be invaluable. They will also minimize transcription errors in the first place. When errors are found, they may be corrected from the original sheet, or it may well be that missing data are irretrievably lost. Missing values can often seriously complicate further data analysis or compromise the power of analysis (e.g., in using ANOVA). Multivariate methods, such as ordination, clustering, discriminant analysis, and multiple regression, are intolerant of missing data. Missing data can be dealt with by the following procedures:

- Either the site or variable with the missing value can be eliminated from the analysis.
- A mean value can be substituted for the missing value. This will ensure that the missing value has minimal weighting in the analysis.
- The missing value can be estimated using spatial or temporal relationships or relationships with other variables.

4.2.3.6 Data analysis and interpretation

The basis for an assessment study is that it is designed to identify a change in the response of the invertebrate community that, in turn, is related to anthropogenic impacts and not associated with normal temporal or spatial variability of the ecosystem. There are three basic approaches to identifying such a change and a number of options for data analysis.

The first approach to identifying change is the traditional method based on contemporary sampling of control sites and potentially impacted sites and an *a posteriori* comparison of the sites, which can be undertaken by either univariate or multivariate statistical analysis. This has been most formally developed in the BACIP (before–after control–impact pairs) designs proposed by Stewart-Oaten et al. (1986, 1992). In this approach, a control and potential impact site are sampled simultaneously over different times, both before and after the initiation of the potential disturbance. Each time period is summarized not by the individual observations at the control and impacted sites, but rather by a measure of the difference between the two sites at the sampling event (i.e., the "pairs"). The BACIP designs require approximate constancy of the difference values. Highly variable difference values mean that contrasts between before and after differences could be expected without an impact having occurred. In addition, independence of the observed differences over time is a further requirement for formal statistical tests.

The second approach is the reference condition concept. This concept was developed in the U.K. by Wright et al. (1984) and has been used in Canada (Reynoldson et al., 1995) and Australia (Norris, 1994). This approach requires the initial acquisition of a large database of reference sites from which a predictive model can be developed to conduct biological assessments at any future sites. These reference databases require information on both invertebrate community structure and habitat attributes, which are used to predict the biological conditions.

The last approach, used most widely in the United States in assessment, is based on additive biological indices (Gerritsen, 1995). These methods have a long history in biological monitoring, beginning with the Saprobic indices to their most recent form as the Index of Biotic Integrity (IBI) developed by Karr and co-workers (Karr, 1991; Karr et al., 1986). The IBI method was first adapted to benthic invertebrates by the Ohio Environmental Protection Agency (1988). The method compares the biological "metrics" at the assessment site to their expected value under regional reference conditions. Such reference conditions are not single cases, such as upstream or paired sites, but are supposed to reflect regional conditions and regional variability under minimal disturbance. The assumption is made that the biological metrics respond at this spatial scale rather than to local conditions.

Whichever of these strategic approaches for study design and definition of impact is used, a number of options for data analysis are available. There are two basic approaches: (1) the use of univariate, inferential statistics, on a suite of variables, either measured or derived, and (2) multivariate methods that search for patterns in a data matrix, usually a species-by-site matrix.

4.2.3.6.1 Univariate methods. The purpose of the analysis will be to detect a significant change between potentially impacted sites and a reference or control site(s) for the variable being examined. When conducting such analyses, it is desirable to distinguish between statistically significant changes and biologically meaningful changes, to have some knowledge of the power of the analysis, to previously establish the amount of change that is required to be detected, and the significance level that is required for both type I and type II errors.

The usual variables analyzed are total abundance (i.e., number of organisms), major taxa abundances and taxa richness (i.e., number of taxa), or derived variables, such as diversity or biotic indices. Each of these variables has some limitations when examined independently.

Measures of abundance are probably the most variable, particularly at the species level, and are dependent on both seasonal variation and annual climate changes. Therefore, some knowledge of normal variation is necessary and methods for normalizing will be required. These measures are probably the least sensitive for detecting change, and only major perturbations will be detectable.

Measures of richness are more likely robust, in that they are, in fact, presence–absence data, and less inherently variable than abundance measures. However, they are affected by the taxonomic effort expended and a consistency of taxonomy is necessary. In reality, the level of taxonomic effort expended is usually related more to the available expertise than to any scientific rationale.

Diversity indices were developed as a way of integrating abundance and richness data. Proponents of diversity indices argue that they measure real ecological properties and processes and are based on ecological theory (e.g., diversity/stability hypothesis [Goodman, 1975] and competitive interaction [Hurlbert, 1971]). The most widely used index is the Shannon–Wiener index (Shannon, 1948). These indices reach their maximum value when all species are evenly distributed, and high diversity is considered a desirable condition. However, such an even distribution of species is not normal in most communities, and therefore the underpinnings have to be questioned. The actual meaning of diversity and how it should be measured are also questionable and have been the subject of much debate (Washington, 1984). Furthermore, different types of pollution may have different impacts in different habitats, and major changes in species composition may occur with no change in diversity. Accordingly, the authors do not consider these measures to be of much value.

The last category of measurements is biotic indices. These are usually developed empirically and have mostly been developed from riverine community studies. They also tend to be regionally specific and have largely evolved from indices developed to respond to pollution by organic contaminants. According to Norris (1995), the biotic indices have been popular for the following reasons: (1) they are seen as useful in condensing data, (2) they are easily understood, (3) they are of more general value than physicochemical measures, (4) they may be less sensitive to methodological differences, and (5) the collection of required data is less expensive than that for inferential statistical analysis. The last few years has seen an increasing use of biological indices for rapid biological assessment (Plafkin et al., 1989; Resh et al., 1995), particularly in the United States, with the purpose of comparing the similarity of a test site to its potential status if it were undisturbed (Karr, 1991).

Indices are usually additive and are either the sum of several measurements or calculated variables. These are termed "metrics," which, in this context, is defined as an ecological attribute of the assemblage sampled and is considered as being responsive to perturbation or disturbance. There are six types of measurements used in these additive indices: richness measures, enumerations, community diversity and similarity measures, biotic indices, functional measures, and combination indices (Barbour et al., 1995; Resh, 1995). A detailed list of the various measures in each of these categories is provided in Table 4-5. Some difficulties with such additive indices have been expressed by Norris (1995). Some of the specific problems that Norris (1995) identifies are as follows:

- Repetitive use of data in the metrics, if two or more indices are incorporated based on the same feature (e.g., number of taxa and Shannon's index), will give undue weight to measure (i.e., the number of taxa) in the final score.
- The dispersion statistics of indices are often ignored when interpreting the data.
- The values are treated as absolute measures that characterize the community which are not subject to sampling error.
- The selection of measures in each measurement category is variable, and there is no theoretical basis for selection which appears to be arbitrary.

Table 4-5 Examples of Measures Used in "Metric" Categories in Deriving Additive Biotic Indices

Richness measures
1. Number of taxa
2. Number of Ephemeroptera, Plecoptera, Trichoptera (EPT) taxa
3. Number of families
4. Niche occupancy forms (Mason, 1979)
5. Number of Ephemeroptera, Plecoptera, Trichoptera, Diptera taxa (considered individually)
6. Number of intolerant snail and mussel taxa
7. Number of Chironomidae
8. Number of Crustacea and Mollusca taxa

Numerative measures
1. Number of individuals (or biomass)
2. % EPT individuals
3. % Chironomidae individuals
4. % Tanytarsini individuals
5. Ratio of EPT/Chironomidae individuals
6. Ratio of Hydropsychidae/ Trichoptera
7. % individuals of numerically dominant taxa
8. % nondipterans
9. % non-Chironomidae, Diptera, and insect individuals
10. Relative abundance of different groups
11. % individuals in dominant taxa
12. Five dominant taxa in common
13. Common taxa index (Shackleford, 1988)
14. Indicator groups (Hayslip, 1992)
15. Relative abundance of different groups
16. % tolerant groups (Ohio EPA, 1987; Yoder and Rankin, 1995)

Diversity and similarity measures
1. Shannon's index (Shannon, 1948)
2. Margalef's index (Margalef, 1951)
3. Menhinick's index (Menhinick, 1964)
4. Simpson's index (Simpson, 1949)
5. Coefficient of community loss (Courtemanch and Davies, 1987)
6. Equitability (Hayslip, 1992)
7. Jaccard coefficient (Jaccard, 1912)
8. Pinkham–Pearson community similarity (Pinkham and Pearson, 1976)
9. Number of dominant taxa in common
10. Number of taxa in common
11. Quantitative similarity index (Barbour et al., 1995)
12. % change in taxa richness
13. Number of unique species per site
14. Missing EPT taxa at site versus reference site
15. Index of community integrity (Hayslip, 1992)

Biotic indices
1. Belgian biotic index (De Pauw and Vanhooren, 1995)
2. Biotic condition index (Plafkin et al., 1989)
3. Biotic index (Chutter, 1972; Hilsenhoff, 1982, 1987, 1988; Lenat, 1993)
4. BMWP score (Wright et al., 1988)
5. Chandler biotic score (Chandler, 1970)
6. Florida index (Ross and Jones, 1979)
7. Indicator organism presence
8. ISO score (International Organization for Standardization, 1984)
9. Community tolerance quotient (Winget and Mangum, 1979)
10. Saprobic index (Zelinka and Marvan, 1961)
11. Dominance of tolerant groups (Plafkin et al., 1989)
12. Indicator assemblage index (Shackleford, 1988; Hayslip, 1992)

Table 4-5 Examples of Measures Used in "Metric" Categories in Deriving Additive Biotic Indices (continued)

Functional measures
1. % shredders (Cummins, 1988)
2. % scrapers (Cummins, 1988)
3. % collector–filterers (Cummins, 1988)
4. % filterers (Cummins, 1988)
5. % predators (Kerans et al., 1992)
6. % omnivores and scavengers (Kerans et al., 1992)
7. Ratio of scrapers/collector–filterers (Cummins, 1988)
8. Ratio of trophic specialists/generalists (Maine Department of Environmental Protection, 1987)
9. Types of functional feeding groups (Cummins, 1988)
10. Functional feeding group similarity (Cummins, 1988)

Combination indices
1. Invertebrate community index (Ohio EPA, 1987)
2. Mean biometric score (Shackleford, 1988)
3. Biological condition score (Winget and Mangum, 1979)

Adapted from Resh et al., 1995.

- One category of measurement, functional feeding groups, is problematic in that the feeding strategies of some species change with instar; therefore the allocation to a specific group is difficult.

The response of indices to features of habitat quality or degradation requires testing, and whether they respond linearly, which is often assumed, or at least monotonically needs to be established. Finally, the behavior of diversity and biotic indices when transformed for univariate statistical analysis (e.g., ANOVA) is not well known because of a lack of published data sets where these measures have been replicated.

4.2.3.6.2 Multivariate methods. Environmental problems generally involve multiple variables, and it is therefore logical that multivariate analytical approaches should be used (Green, 1979; Norris, 1995). The insights that can be gained from univariate approaches, such as ANOVA and regression analyses, are limited. A composite variable, such as total abundance, when analyzed univariately, may hide different but compensating trends in the abundance of different taxa (Green, 1979; Gauch, 1982). Derived biological measures, such as many of those identified in Table 4-4, can be ambiguous indicators of change as they do not always respond monotonically along pollution gradients (Bernstein and Smith, 1986). While the use of indicator species is theoretically superior to derived measures of their natural variability, this can be difficult to interpret when single-species data are examined univariately. Multivariate procedures consider each species (i.e., taxa) as a variable, and either the presence/absence or abundance of each species is an attribute of a site or time. Subtle changes in the species composition or abundance of a single species across sites are not masked by the need to summarize the combined attributes of a site into a single value. Such methods should therefore be more appropriate for detecting and understanding spatial and temporal changes in benthic fauna. Furthermore, multivariate methods also allow the simultaneous examination of spatial and temporal trends in benthic communities with multiple environmental variables, and therefore the relative weighting of natural gradients can be distinguished from those associated with pollution gradients (Clarke and Ainsworth, 1993). Finally, these methods allow the development of predictive models of species composition that can be used for assessment purposes (Wright et al., 1984; Reynoldson et al., 1995).

While it is not possible to discuss here all the intricacies of the methods used, the basic steps involved are outlined and some of the strategies are presented. Many multivariate methods, and the association measures on which they are based, have now been rigorously tested using simulated data with known characteristics, which may be superior to the use of field data where the variable attributes are unknown. Association measures have been evaluated by Faith et al. (1987) and Jackson (1993), ordination methods by Minchin (1987) and Kenkel and Orloci (1986), and classification methods by Belbin and McDonald (1993). The first stage in conducting multivariate analyses is the assembly of the original data matrix. This will usually be a site-by-species square matrix, and, ideally, there will be a matching site by a matrix of environmental variables.

Field et al. (1982) outlined a strategy for the analysis of data on community structure data in the form of an abundance matrix where columns represent species and rows are separate samples. As indicated earlier, multivariate methods are intolerant of missing data, and if a cell in the matrix has a missing value, then either the sampling event or the variable must be eliminated from the matrix or a value inserted. Once the matrix is assembled, usually in one of the common spreadsheet software packages (for example, Lotus 123, Excel, or Quattro Pro), it can be imported into the data analysis package of choice. In the approach described by Field et al. (1982), the following steps have been identified:

1. The biotic relationship between any two samples is converted into a coefficient that measures similarity (or dissimilarity) in species composition. At this juncture, a decision is necessary on which association metric to use to measure similarity between each pair of samples. Ideally, any metric used should have a fairly robust and monotonic and linear relationship with ecological distance. Faith et al. (1987) found the Kulczynski, Bray–Curtis, and relativized Manhattan metrics to be more robust than Chord distance, Kendall's coefficient, chi-squared, Manhattan distance, or Euclidean distance. Jackson (1993)

found Bray–Curtis and the correlation coefficient to provide more consistent solutions than Euclidean distance or the covariance matrix when different data standardizations were used.

2. The resulting triangular matrix of similarities between each pair of samples is used to classify the samples into groups, by hierarchical agglomerative clustering with group-average linking (UPGMA), and/or to plot the sample interrelationships in ordination space using nonmetric multidimensional scaling.
3. Relationships between the species can be examined by transposing the data matrix and repeating the classification and ordination from a similarity matrix computed between species pairs. This allows the significance of species to the groups formed by classification or the ordination scores to be established.
4. The relationship between the community data and matching environmental data can be examined by superimposing values for each abiotic variable onto the biotic ordination.

This strategy or variations of it have been successfully used in a number of marine (Bayne et al., 1988; Addison and Clarke, 1990; Warwick and Clarke, 1991) and freshwater studies (Wright et al., 1984; Corkum and Currie, 1987; Ormerod and Edwards, 1987; Faith and Norris, 1989; Reynoldson et al., 1995).

Clarke (1993) identifies some problems with the mechanics recommended in Field's strategy (e.g., the selection of the I statistic for identifying indicator species and the superimposition of environmental data variable by variable). However, the general strategy is supported. In his paper, Clarke (1993) addresses four basic questions that will affect study design and interpretation:

1. What community attributes are to be measured?
2. Are data to be standardized?
3. Should data be transformed? Is the emphasis to be on common species (untransformed) or rare species (transformed)?
4. Which association measure should be used?

Clarke (1993) modified Field's strategy somewhat and proposed a framework for nonparametric multivariate analysis of community data that is rooted in the among-sample similarity matrix and where inferences are drawn from the ranking in the matrix. The components of Clarke's version are the following:

1. *The display of community patterns through ordination and clustering*: Clarke (1993) recommends nonmetric multidimensional scaling or correspondence analysis (nonrobust behavior of χ^2 distance measure) rather than PCA (because of its assumptions of linearity).

2. *The determination of species responsible for sample groupings observed from a cluster analysis*: This can be carried out using the methods outlined by Clarke (1993) based on the contribution each species makes to the similarity coefficient or using the principal axis correlation (Faith and Norris, 1989; Belbin, 1992). The statistical significance of the correlations can be tested through a Monte Carlo procedure for each species.
3. *Testing for spatial and temporal differences in the measured community structure when samples are replicated and an a priori hypothesis defined*: The hypothesis of no-site to site differences can be tested using the analysis of similarities (ANOSIM) procedure (Clarke and Green, 1988).
4. *Linking biological groupings to environmental variables when a set of abiotic data has been collected to match each biotic sample*: This step is based on the premise that sites which are similar in their environmental variables will be similar in their species composition, if the relevant environmental variables have been measured. If this is the case, separate ordinations of biotic and abiotic data would be expected to show a good match. Clarke and Ainsworth (1993) have developed a simple optimization routine (BIO-ENV procedure) which selects the subset of environmental variables that maximizes a rank correlation between the biotic and abiotic similarity matrices. The best match can be quantitatively established using procrustees analysis (Gower, 1971). An alternate approach is to again use the principal axis correlation.

The utility of these multivariate methods has been demonstrated in a number of applications, particularly in pollution studies, where they have been shown to be sensitive in detecting community change (Clarke, 1993) and robust to a substantial degree of taxonomic aggregation (Warwick, 1993). They have been more widely used in marine studies of fine-grained sediment (examples in Clarke, 1993; Warwick, 1993; Warwick and Clarke, 1993), but have also been used in freshwater sediment studies (Reynoldson et al., 1995) and in freshwater stream studies (Faith and Norris, 1989). The methods are well suited to some common features of soft-sediment data, such as the large species sets (i.e., usually greater than the number of samples) and the sparse and skewed abundance matrices, because of the spatial heterogeneity and aggregate distribution patterns of most benthic organisms. These nonparametric methods lack some of the sophistication of other multivariate methods. However, this is compensated for by their widespread validity and the comparative ease with which they can be understood. This is a major advantage when communicating results from the large data sets that are developed by assessment studies.

4.3 Recommended approaches and methods

4.3.1 Guide to material

The information provided in this chapter is an attempt to synthesize current thought on biological sediment assessment. The rapid evolution of ideas in this field is illustrated by the large number of recent publications; for example, in recent years, six books were published in this area: Burton (1992), Peakall (1992), Calow (1993b), Rosenberg and Resh (1993), Davis and Simon (1995), and Hoffman et al. (1995). Therefore, the authors cannot possibly cover the entire field in the limited number of pages of this chapter. Rather, they have attempted to briefly cover the entire field and provide the reader with access to other relevant literature. In the tables in particular, the reader is led to other, more detailed reviews that will provide access to more comprehensive coverage. The authors have attempted to provide advice on the questions and considerations of which the field scientist needs to be aware and, where they can, have given their own personal views on the advantages and disadvantages of various methods from a pragmatic operational view.

4.3.1.1 How to select methods

When selecting methods, the objectives of the study must be borne in mind. A number of different studies were outlined in the previous chapter of this book, all of which would require a different set of methods to be used:

1. Screening for horizontal and vertical gradients of sediment toxicity in an area or region for a general classification
2. Measuring efficiency of remediation programs
3. Measuring impacts of dredging and channelization works that provide a resuspension of the sediment into the water column
4. Regulating hazards of effluent discharges
5. Assessing bioavailability and hazards of chemicals and establishing numerical targets for quality standards and criteria of sediment toxicity
6. Bioaccumulation of toxicants in tissues of aquatic organisms for protection of the ecosystem and health
7. Establishing a causality between community data and contaminants

However, when selecting methods to use, the authors would suggest the following general principles be borne in mind:

- Both chemical and biological information are essential in any well-designed study. Biological data provide the "so what" information. As organisms ourselves, it is the impairment of the biological component of the ecosystem with which enlightened self-interest should be concerned. Chemical information is the key to identifying "cause" and allows focus on remedial actions.

- There is no such thing as the universal test, method, or species, and multiple lines of evidence are optimal.
- Conversely, it is neither possible nor desirable to measure everything everywhere.
- In selecting methods, the user should assess them against the five principles described above: *relevance, robustness, methodology, sensitivity,* and *appropriateness/application*. The authors have attempted to make such an assessment in a general sense in Tables 4-2 and 4-3. However, these rankings are very subjective, and local circumstances may substantially affect any of these categories.

4.4 Summary

A number of approaches using either or both chemical and biological measures have been used to assess sediment contamination. Chemical analyses are most frequently performed on bulk sediment for total concentrations of contaminants, such as total PCBs, total PAHs, etc. Analyses are less frequently conducted on pore or interstitial water, and differentiation of chemical forms of contaminants is infrequently performed either in bulk sediments or interstitial water. To assess the extent of contamination and the need for remedial action, chemical concentrations are compared to background (i.e., prehistoric concentrations, predetermined objectives, criteria, or standards for water or sediment quality) or to levels known to produce biological effects. The various methods used to assess sediment contamination include the equilibrium partitioning approach, the screening level concentration approach, and the apparent effects threshold (Persaud et al., 1992; Zarull and Reynoldson, 1992). While each of these approaches to developing sediment objectives or criteria has its own advantages and disadvantages (Giesy and Hoke, 1990), they collectively suffer from the need to infer biological impact rather than direct measurements of biological effect (Reynoldson and Zarull, 1993). Although most of these approaches will provide a high degree of protection to the aquatic ecosystem, as part of a comprehensive remedial action plan for nearshore areas they are unlikely to prove adequate justification for the expensive removal of extensive volumes of sediments, the treatment of in-place pollutants, or other methods of contaminated sediment mitigation. It has been shown by Painter (1992) that many sediments in areas of the Great Lakes which are well removed from source(s) of contamination contain concentrations of priority pollutants which exceed both the low and severe effects concentrations for aquatic biota. Most municipal, provincial, or federal budgets do not have the capacity for the extensive remedial action which would be required to deal with this amount of sediment, nor is there necessarily a need for such a degree of remedial action.

As the fundamental reason for the development of guidelines, objectives, or criteria for sediments is to provide and protect a sustainable and reproducing aquatic biota, alternative methods for developing such numerical data using biological assessments are needed. Biological assessments usually employ either structural or functional measures of impact or both. In the case of sediments, the authors consider benthic invertebrates as the most appropriate group of organisms to use. Their populations are relatively stable in time, with life cycles of one or more years, and their taxonomy can be determined at least to the level of genus and, in some cases, species. In addition, their response(s) to environmental change has been extensively studied.

Several authors have proposed the use of benthic invertebrates in the assessment and management of contaminated sediments. For example, Chapman and co-workers (Chapman and Long, 1983; Long and Chapman, 1985; Chapman, 1986) have described the sediment triad approach, which incorporates aspects of sediment chemistry, toxicity testing, and analysis of the benthic community structure. The U.S.A.–Canada International Joint Commission has suggested a sediment management strategy that incorporates both assessment and remediation (International Joint Commission, 1987, 1988). The United States has embarked on a program to examine various approaches to sediment assessment and remediation (i.e., ARCS program). Canada is also addressing these issues via the Great Lakes Action Plan. However, there are still major objections to the employment of the structure of the benthic community to set sediment criteria or objectives. The main criticisms are their lack of universality (i.e., they are, of necessity, site specific) and the inability of researchers to establish quantitative objectives for their application (i.e., what should the community "look" like). Universal guidelines may not be possible due to the very nature and complexity of sediment–contaminant–biota relationships and the diversification of biological and geological components over a large regional area.

Recent developments in the analysis of biological data using multivariate statistical techniques have shown extremely promising results in interpreting changes in community structure based on simple environmental parameters. As a first step toward the identification of the best achievable community for a specific type of habitat, there is a need to define reference communities based on chemical, physical, geological, and geographical features in areas free from contamination. The greatest effort in this direction has been made in the U.K. (Wright et al., 1984; Armitage et al., 1987; Moss et al., 1987). However, similar studies have been conducted in North America (Corkum and Currie, 1987), continental Europe (Johnson and Wielderholm, 1989), and South Africa (King, 1981). An extensive collection of data at nonpolluted sites in the U.K. has resulted in the identification of a number

Table 4-6 Ability to Predict Benthic Community Invertebrate Structure

Habitat	No. sites	No. community groups	Predictor variables	Prediction success rate	Reference
Rivers	286	16	28	76.1	Wright et al., 1984
Streams	79	5	26	70.9	Corkum and Currie, 1987
Rivers	45	6	5	68.9	Ormerod and Edwards, 1987
Rivers	54	5	9	79.6	King, 1981
Lakes	68	7	11	90.0	Johnson and Wiederholm, 1989
Rivers	35	5	6	75.4	Reynoldson and Zarull, 1993
Lakes	95	6	9	90.0	Reynoldson et al., 1995

of natural communities (Wright et al., 1984). Environmental variables, such as latitude, substrate type, temperature, and depth, were used to correctly predict the benthic communities at 75% of 268 sites. The observed community was similar to either the predicted or the next most similar predicted community. At lower levels of community detail, even greater accuracy of prediction was observed. Other studies have shown a similar predictive capability (see Table 4-6), and the accuracy of prediction in these studies ranges from 68 to 90%, in habitats varying from large lakes to small streams on three continents. In other words, based on a few physical variables, the expected community assemblage(s) at a location can be defined from a predictive model. Such predicted communities, or key species in the communities, can be used to establish site-specific guidelines, which may be compared with the actual species composition. Thus, determination can be made as to whether or not the guideline is being met.

To date, this approach has only been used on community assemblages with species as the classifying variables. There is no reason why the same approach cannot be used with both structural and functional variables, such as reproduction or growth or biomarker tests that can be included in a multivariate design to classify both predictable communities and responses in tests.

References

Addison, R.F. and Clarke, K.R., Eds., Biological effects of pollutants in a subtropical environment, *J. Exp. Mar. Biol. Ecol.*, 138, 1, 1990.

Armitage, P.D., Gunn, R.J.M., Furse, M.T., Wright, J.F., and Moss, D., The use of prediction to assess macroinvertebrate response to river regulation, *Hydrobiologia*, 144, 25, 1987.

Barbour, M.T., Stribling, J.B., and Karr, J.R., Multimetric approach for establishing biocriteria and measuring biological condition, in *Biological Assessment and Criteria: Tools for Water Resource Planning and Decision Making*, Davis, W.S. and Simon, T., Eds., Lewis Publishers, Boca Raton, FL, 1995, 63.

Baudo, R., Giesy, J.P., and Muntau, H., Eds., *Sediments: Chemistry and Toxicity of In-Place Pollutants,* Lewis Publishers, Boca Raton, FL, 1990, 405.

Bayne, B.L., Aspects of the metabolism of *Mytilus edulis* during starvation, *Neth. J. Sea Res.,* 7, 399, 1973a.

Bayne, B.L., Physiological changes in *Mytilus edulis* L. induced by temperature and nutritive stress, *J. Mar. Biol. Assoc. U.K.,* 53, 39, 1973b.

Bayne, B.L., Clarke, K.R., and Grey, J.S., Eds., Biological effects of pollutants: results of a practical workshop, *Mar. Ecol. Prog. Ser.,* 46, 1, 1988.

Belbin, L., PATN, Technical reference, Division of Wildlife and Ecology, C.S.I.R.O., Canberra, Australia, 1992.

Belbin, L. and McDonald, C., Comparing three classification strategies for use in ecology, *J. Veg. Sci.,* 4, 341, 1993.

Benson, W. and Di Giulio, R., Biomarkers in hazard assessments of contaminated sediments, in *Sediment Toxicity Assessment,* Burton, G.A., Jr., Ed., Lewis Publishers, Boca Raton, FL, 1992, 241.

Benson, W.H., Baer, K.N., and Watson, C.F., Metallothionein as a biomarker of environmental metal contamination, in *Biomarkers of Environmental Contamination,* McCarthy, J.F. and Shugart, L.R., Eds., Lewis Publishers, Boca Raton, FL, 1990.

Bernstein, B.B. and Smith, R.W., Community approaches to monitoring, in *Oceans '86 Conference Record: Science–Engineering Adventure,* Proc. of a Conference Hosted by the Washington, D.C. Section of the Marine Technology Soc., Institute of Electrical and Electronic Engineering, Piscataway, NJ, 1986, 934.

Bickham, J.W., Flow cytometry as a technique to monitor the effects of environmental genotoxins on wildlife populations, in *In Situ Evaluation of Biological Hazards of Environmental Pollutants,* Sandhu, S., Lower, W.R., DeSerres, F.J., Suka, W.A., and Tice, R.R., Eds., Environmental Research Series, Vol. 38, Plenum Press, New York, 1990, 97.

Black, J., Field and laboratory studies of environmental carcinogenesis in Niagara River fish, *J. Great Lakes Res.,* 9, 326, 1983.

Brinkhurst, R.O., Hamilton, A.L., and Herrington, H.B., Components of the Bottom Fauna of the St. Lawrence, Great Lakes, No. PR 33, Great Lakes Institute, University of Toronto, Toronto, 1968.

Bry, C., Daily variations in plasma cortisol levels of individual female rainbow trout *Salmo gairdneri*: evidence for a post-feeding peak in well-adapted fish, *Gen. Comp. Endocrinol.,* 48, 462, 1982.

Bulow, F.J., RNA–DNA ratios as indicators of growth in fish: a review, in *Age and Growth of Fish,* Summerfelt, R.C. and Hall, G.E., Eds., Iowa State University Press, Ames, 1987, 45.

Burgess, R.M. and Scott, K.J., The significance of in place contaminated marine sediments on the water column: processes and effects, in *Sediment Toxicity Assessment,* Burton, G.A., Jr., Ed., Lewis Publishers, Boca Raton, FL, 1992, 129.

Burke, M.D. and Mayer, R.T., Ethoxyresorufin: direct fluorometric assay of microsomal O-dealkylation which is preferentially inducible by 3-methylcholanthrene, *Drug Metab. Disp.,* 2, 583, 1974.

Burke, M.D., Thompson, S., Elcombe, C.R., Halpert, J., Haaparanta, T., and Mayer, R.T., Etoxy, pentoxy- and benzyloxyphenoxazones and homologues: a series of substarts to distinguish between different induced cytochromes P-450, *Biochem. Pharmacol.,* 34, 3337, 1985.

Burton, G.A., Jr., *Sediment Toxicity Assessment*, Lewis Publishers, Boca Raton, FL, 1992, 457.

Burton, G.A., Jr. and Stemmer, B.L., Evaluation of surrogate tests in toxicant impact assessments, *Toxicol. Assess.*, 3, 225, 1988.

Burton, G.A., Jr., Drotar, A., Lazorchak, J.M., and Bahls, L.L., Relationship of microbial activity and *Ceriodaphnia* responses to mining impacts on the Clark Fork River, Montana, *Arch. Environ. Contam. Toxicol.*, 16, 523, 1987.

Burton, G.A., Jr., Stemmer, B.L., Winks, K.L., Ross, P.E., and Burnett, L.C., A multitrophic level evaluation of sediment toxicity in Waukegan and Indiana Harbors, *Environ. Toxicol. Chem.*, 8, 1057, 1989.

Cairns, J., Niederlehner, B.R., and Smith, E.P., The emergence of functional attributes as endpoints in ecotoxicology, in *Sediment Toxicity Assessment*, Burton, G.A., Jr., Ed., Lewis Publishers, Boca Raton, FL, 1992, 111.

Calow, P., General principles and overview, in *Handbook of Ecotoxicology*, Vol. 1, Calow, P., Ed., Blackwell Scientific, London, 1993a, 1.

Calow, P., Ed., *Handbook of Ecotoxicology*, Vol. 1, Blackwell Scientific, London, 1993b.

Chandler, J.R., A biological approach to water quality management, *Water Pollut. Control*, 69, 415, 1970.

Chapman, P.M., Sediment quality criteria from the sediment quality triad: an example, *Environ. Toxicol. Chem.*, 5, 957, 1986.

Chapman, P.M. and Long, E.R., The use of bioassays as part of a comprehensive approach to marine pollution assessment, *Mar. Pollut. Bull.*, 14, 81, 1983.

Chapman, P.M., Power, E.A., and Burton, G.A., Jr., Integrative assessments in aquatic ecosystems, in *Sediment Toxicity Assessment*, Burton, G.A., Jr., Ed., Lewis Publishers, Boca Raton, FL, 1992, 313.

Chaudry, H.S., Nickel toxicity on carbohydrate metabolism of a freshwater fish *Colisa fasciatus*, *Toxicol. Appl. Pharmacol.*, 50, 241, 1984.

Chutter, F.M., A re-appraisal of Needham and Usinger's data on the variability of a stream fauna when sampled with a Surber sampler, *Limnol. Oceanogr.*, 17, 139, 1972.

Clarke, K.R., Non-parametric multivariate analyses of changes in community structure, *Aust. J. Ecol.*, 18, 117, 1993.

Clarke, K.R. and Ainsworth, M., A method linking multivariate community structure to environmental variables, *Mar. Ecol. Prog. Ser.*, 92, 205, 1993.

Clarke, K.R. and Green, R.H., Statistical design and analysis for a "biological effects" study, *Mar. Ecol. Prog. Ser.*, 46, 213, 1988.

Coler, R.A., Coler, M.S., and Kostecki, P.T., Tubificid behaviour as a stress indicator, *Water Res.*, 22, 263, 1988.

Corkum, L.D. and Currie, D.C., Distributional patterns of immature Simuliidae (Diptera) in northwestern North America, *Freshwater Biol.*, 17, 201, 1987.

Courtemanch, D.L., Trophic Classification of Maine Lakes Using Benthic Chironomid Fauna, unpublished manuscript, 1989.

Courtemanch, D.L. and Davies, S.P., A coefficient of community loss to assess detrimental change in aquatic communities, *Water Res.*, 21, 217, 1987.

Cummins, K.W., Rapid bioassessment using functional analysis of running water invertebrates, in Proc. First National Workshop on Biological Criteria, Simon, T.P., Holst, L.L., and Shepard, L.J., Eds., U.S. Environmental Protection Agency, Chicago, 1988, 49.

Davis, W.S. and Simon, T., *Biological Assessment and Criteria: Tools for Water Resource Planning and Decision Making*, Lewis Publishers, Boca Raton, FL, 1995, 415.

Dawe, C.J., Stanton, M.F., and Shwartz, F.J., Hepatic neoplasms in native bottom-feeding fish of Deep Creek Lake, Maryland, *Cancer Res.*, 24, 1194, 1964.

De Pauw, N. and Vanhooren, G., Method for biological quality assessment of watercourses in Belgium, *Hydrobiologia*, 100, 153, 1995.

Detra, R.L. and Collins, W.J., The relationship of parathion concentration, exposure time, cholinesterase inhibition and symptoms of toxicity in midge larvae (Chironomidae: Diptera), *Environ. Toxicol. Chem.*, 10, 1089, 1991.

Dey, A.C., Kiceniuk, J.W., Williams, U.P., Khan, R.A., and Payne, J.F., Long term exposure of marine fish to crude petroleum. I. Studies on liver lipids and fatty acids in cod (*Gadus morhua*) and winter flounder (*Pseudopleuronectes americanus*), *Comp. Biochem. Physiol.*, 75C, 93, 1983.

Diaz, R.J., Ecosystem assessment using estuarine and marine benthic community structure, in *Sediment Toxicity Assessment*, Burton, G.A., Jr., Ed., Lewis Publishers, Boca Raton, FL, 1992, 67.

Di Giulio, R.T., Indices of oxidative stress as biomarkers for environmental contamination, in *Aquatic Toxicology and Risk Assessment*, Vol. 14, Mayes, M.A. and Barron, M.G., Eds., American Society for Testing and Materials, Philadelphia, 1991, 15.

Di Giulio, R.T., Washburn, P.C., Wenning, R.J., Winston, G.W., and Jewell, C.S., Biochemical responses in aquatic animals: a review of determinants of oxidative stress, *Environ. Toxicol. Chem.*, 8, 1103, 1989.

Dillon, T.M. and Benson, W.H., Effects of PCB-contaminated sediments on reproductive success of fathead minnows: relationship between tissue residues and biological effects, paper presented at 8th Annual Meeting of the Society of Environmental Toxicology and Chemistry, Pensacola, FL, 1987.

DiMichelle, L. and Taylor, M.H., Histopathological and physiological responses of *Fundulus heteroclitus* to napthalene exposure, *J. Fish. Res. Board Can.*, 35, 1060, 1978.

Donaldson, E.M. and Dye, H.M., Corticosteroid concentrations in sockeye salmon (*Oncorhynchus nerka*) exposed to low concentrations of copper, *J. Fish. Res. Board Can.*, 32, 533, 1975.

Downing, J.A., Sampling the benthos of standing waters, in *A Manual on Methods for the Assessment of Secondary Productivity in Fresh Waters*, 2nd ed., Downing, J.A. and Rigler, F.H., Eds., IBP Handbook 17, Blackwell Scientific, Oxford, England, 1984, 87.

Dunn, B., Black, J., and Maccubin, A., ^{32}P-postlabeling analysis of aromatic DNA adducts in fish from polluted areas, *Cancer Res.*, 47, 6543, 1987.

Dutton, M.D., Stephenson, M., and Klaverkamp, J.F., A mercury saturation assay for measuring metallothionein in fish, *Environ. Toxicol. Chem.*, 12, 1193, 1993.

Edwards, C.A. and Fisher, S.W., The use of cholinesterase measurements in assessing the impacts of pesticides on terrestrial and aquatic invertebrates, in *Cholinesterase Inhibiting Insecticides: Their Impact on Wildlife and the Environment*, Mineau, P., Ed., Elsevier Science, Amsterdam, 1991, 348.

Elliott, J.M., *Some Methods for the Statistical Analysis of Samples of Benthic Invertebrates*, 2nd ed., Freshwater Biological Association Scientific Publ. 25, 1977, 156.

Elliott, J.M. and Tullet, P.A., *A Bibliography of Samplers for Benthic Invertebrates*, Freshwater Biological Association Occasional Publ. 4, 1978, 1.

Engel, D.W. and Brouwer, M., Metallothionein and metallothionein-like proteins: physiological importance, *Adv. Comp. Environ. Physiol.*, 4, 53, 1989.

Fabacher, D.L. and Baumann, P.C., Enlarged livers and hepatic microsomal mixed-function oxidase components in tumor-bearing brown bullheads from a chemically contaminated river, *Environ. Toxicol. Chem.*, 4, 703, 1985.

Fairbrother, A., Marden, B.T., Bennett, J.K., and Hooper, M.J., Methods used in the determination of cholinesterase activity, in *Cholinesterase Inhibiting Insecticides: Their Impact on Wildlife and the Environment*, Mineau, P., Ed., Elsevier Science, Amsterdam, 1991, 348.

Faith, D.P. and Norris, R.H., Correlation of environmental variables with patterns of distribution and abundance of common and rare freshwater macroinvertebrates, *Biol. Conserv.*, 50, 77, 1989.

Faith, D.P., Minchin, P.R., and Belbin, L., Compositional dissimilarity as a robust measure of ecological distance, *Vegetatio*, 69, 57, 1987.

Field, J.G., Clarke, K.R., and Warwick, R.M., A practical strategy for analysing multispecies distribution patterns, *Mar. Ecol. Prog. Ser.*, 8, 37, 1982.

Fletcher, D.J., Plasma glucose and plasma fatty acid levels of *Limanda limanda* (L.) in relation to season, stress, glucose loads and nutritional state, *J. Fish. Biol.*, 25, 629, 1984.

Gahnstrom, G., Sediment oxygen uptake in the acidified Lake Gardsjon, Sweden, *Ecol. Bull.*, 37, 276, 1985.

Galen, R.S., Multiphasic screening and biochemical profiles: state of the art, in *Progress in Clinical Pathology*, Vol. 6, Stefanini, M. and Isenberg, H.D., Eds., Grune and Stratton, New York, 1975, 83.

Gardner, D. and Riley, P.J., The component fatty acids of the lipids of some species of marine and freshwater molluscs, *J. Mar. Biol. Assoc. U.K.*, 52, 827, 1972.

Gauch, H.G., Jr., *Multivariate Analysis in Community Ecology*, Cambridge University Press, Cambridge, England, 1982, 289.

Gerritsen, J., Additive biological indices for resource management, *J. North Am. Benthol. Soc.*, 14, 451, 1995.

Gibson, R.F., Ludke, J.L., and Ferguson, D.E., Sources of error in the use of fish brain acetylcholinesterase as a monitor for pollution, *Bull. Environ. Contam. Toxicol.*, 4, 17, 1969.

Giesy, J.P. and Hoke, R.A., Freshwater sediment toxicity bioassessment: rationale for species selection and test design, *J. Great Lakes Res.*, 15, 539, 1989.

Giesy, J.P. and Hoke, R.A., Freshwater sediment quality criteria: toxicity bioassessment, in *Sediments: Chemistry and Toxicity of In-Place Pollutants*, Baudo, R., Giesy, J.P., and Muntau, H., Eds., Lewis Publishers, Boca Raton, FL, 1990, 265.

Gill, T.S. and Pant, J.C., Cadmium toxicity: inducement of changes in blood and tissue metabolites in fish, *Toxicol. Lett.*, 18, 195, 1983.

Goksoyr, A., A semi-quantitative cytochrome P450IA1 ELISA: a simple method for studying the monooxygenase induction response in environmental monitoring and ecotoxicological testing of fish, *Sci. Total Environ.*, 101, 255, 1991.

Goodman, D., The theory of diversity–stability relationships in ecology, *Q. Rev. Biol.*, 50, 237, 1975.

Gower, J.C., Statistical methods of comparing different multivariate analyses of the same data, in *Mathematics in the Archaeological and Historical Sciences*, Hodson, F.R., Kendall, D.G., and Tautu, P., Eds., Edinburgh University Press, Edinburgh, 1971, 138.

Green, R.H., *Sampling Design and Statistical Methods for Environmental Biologists*, John Wiley, New York, 1979, 257.

Hadderingh, R.H., van der Velde, G., and Schnabel, P.G., The effects of heated effluent on the occurrence and reproduction of the freshwater limpets *Ancylus fluviatus* (Müller, 1774), *Ferrissia wautieri* (Mirolli, 1960) and *Acroloxus lacustris* (L., 1758) in two Dutch waterbodies, *Hydrobiol. Bull.*, 21, 193, 1987.

Hamer, D.H., Metallothionein, *Annu. Rev. Biochem.*, 55, 913, 1986.

Hayslip, G.A., EPA Region 10 In-Stream Biological Monitoring Handbook (for Wadable Streams in the Pacific Northwest), U.S. EPA-Region 10, Environmental Service Division, Seattle, 1992.

Heinis, F., Timmermans, K.R., and Swain, W.R., Short-term sublethal effects of cadmium on the filter feeding chironomid larva *Glypotendipes pallens* (Meigen) (Diptera), *Aquat. Toxicol.*, 16, 73, 1990.

Hellawell, J.M., *Biological Indicators of Freshwater Pollution and Environmental Management*, Elsevier Science, London, 1986, 546.

Henry, M.G., Chester, D.N., and Mauck, W.L., Role of artificial burrows in *Hexagenia* toxicity tests: recommendations for protocol development, *Environ. Toxicol. Chem.*, 5, 553, 1986.

Hilsenhoff, W.L., Use of Arthropods to Evaluate Water Quality of Streams, Technical Bull. No. 100, Wisconsin Department of Natural Resources, Madison, 1977.

Hilsenhoff, W.L., Using a Biotic Index to Evaluate Water Quality in Streams, Technical Bull. No. 132, Wisconsin Department of Natural Resources, Madison, 1982.

Hilsenhoff, W.L., An improved biotic index of organic stream pollution, *Great Lakes Entomol.*, 20, 31, 1987.

Hilsenhoff, W.L., Rapid field assessment of organic pollution with a family-level biotic index, *J. North Am. Benthol. Soc.*, 7, 65, 1988.

Hinton, D.E., Baumann, P.C., Gardner, G.R., Hawkins, W.E., Hendricks, J.D., Murchelano, R.A., and Okihiro, M.S., Histopathological biomarkers, in *Biomarkers: Biochemical, Physiological, and Histopathological Markers of Anthropogenic Stress*, Hugget, R.J., Kimerle, R.A., Mehrle, P.M., and Bergmen, H.L., Eds., Lewis Publishers, Boca Raton, FL, 1992, 155.

Hoffman, D.J., Rattner, B.A., Burton, G.A., Jr., and Cairns, J., Jr., *Handbook of Ecotoxicology*, Lewis Publishers, Boca Raton, FL, 1995, 755.

Hugget, R.J., Kimerle, R.A., Mehrle, P.M., and Bergmen, H.L., Eds., *Biomarkers: Biochemical, Physiological and Histopathological Markers of Anthropogenic Stress*, Lewis Publishers, Boca Raton, FL, 1992, 347.

Hurlbert, S.H., The nonconcept of species diversity: a critique and alternative parameters, *Ecology*, 52, 577, 1971.

International Joint Commission, Guidance on the Characterization of Toxic Substances Problems in Areas of Concern in the Great Lakes Basin, Report to the Great Lakes Water Quality Board, Windsor, Ontario, 1987, 179.

International Joint Commission, Procedures for the Assessment of Contaminated Sediment Problems in the Great Lakes, Report to the Great Lakes Water Quality Board, Windsor, Ontario, 1988, 140.

International Organization for Standardization, Water Quality — Assessment of the Water Quality and Habitat Quality of Rivers by a Microinvertebrate "Score," Draft proposal ISO/DP 8689, 1984.

Jaccard, P., The distribution of flora in the alpine zone, *New Phytol.*, 11, 37, 1912.

Jackson, D.A., Multivariate analysis of benthic invertebrate communities: the implication of choosing particular data standardizations, measure of association and ordination methods, *Hydrobiologia*, 268, 9, 1993.

Johnson, R.K. and Wiederholm, T., Classification and ordination of profundal macroinvertebrate communities in nutrient poor, oligo-mesohumic lakes in relation to environmental data, *Freshwater Biol.*, 21, 375, 1989.

Johnson, R.K., Wiederholm, T., and Rosenberg, D.M., Freshwater biomonitoring using individual organisms, populations and species assemblages of benthic macroinvertebrates, in *Freshwater Biomonitoring and Benthic Invertebrates*, Rosenberg, D.M. and Resh, V.H., Eds., Chapman and Hall, New York, 1993, 40.

Jones, R.D. and Hood, N.A., The effects of organophosphorus pesticides on estuarine ammonia oxidizers, *Can. J. Microbiol.*, 56, 2488, 1980.

Jones, W.G. and Walker, K.F., Accumulation of iron, manganese, zinc and cadmium by the Australian freshwater mussel *Velesunio ambiguus* (Phillipi) and its potential as a biological monitor, *Aust. J. Mar. Freshwater Res.*, 30, 741, 1979.

Karr, J.R., Biological integrity: a long neglected aspect of water resource management, *Ecol. Appl.*, 1, 66, 1991.

Karr, J.R., Fausch, K.D.F., Angermeier, P.L., Yant, P.R., and Schlosser, I.J., Assessing Biological Integrity in Running Waters: A Method and Its Rationale, Illinois Natural History Survey, Special Publ. No. 5, 1986, 1-28.

Kearns, P.K. and Atchison, G.J., Effects of trace metals on growth of yellow perch (*Perca flavescens*) as measured by RNA–DNA ratios, *Environ. Biol. Fish.*, 4, 383, 1979.

Keilty, T.J. and Landrum, P.F., Population-specific toxicity responses by the freshwater Oligochete, *Stylodilus heringianus*, in natural Lake Michigan sediments, *Environ. Toxicol. Chem.*, 9, 1147, 1990.

Kenkel, N.C. and Orloci, L., Applying metric and non-metric multidimensional scaling to ecological studies: some new results, *Ecology*, 67, 919, 1986.

Kennedy, S.W., Wigfield, D.C., and Fox, G.A., Tissue porphyrin pattern determination by high speed high performance liquid chromatography, *Anal. Biochem.*, 157, 1, 1986.

Kerans, B.L., Karr, J.R., and Ahlstedt, S.A., Aquatic invertebrate assemblages: spatial and temporal differences among sampling protocols, *J. North Am. Benthol. Soc.*, 11, 377, 1992.

King, J.M., The distribution of invertebrate communities in a small South African river, *Hydrobiologia*, 83, 43, 1981.

Klerks, P.L. and Levinton, J.S., Rapid evolution of metal resistance in a benthic oligochaete inhabiting a metal polluted site, *Biol. Bull.*, 176, 135, 1989.

Klotz, A.V., Stegeman, J.J., and Walsh, C., An alternative 7-ethoxyresorufin O-deethylase activity assay: a continuous visible spectrophotometric method for measurement of cytochrome P-450 monooxygenase activity, *Anal. Biochem.*, 140, 138, 1984.

La Point, T.W., Signs and measurements of ecotoxicity in the aquatic environment, in *Handbook of Ecotoxicology*, Hoffman, D.J., Rattner, B.A., Burton, G.A., Jr., and Cairns, J., Jr., Eds., Lewis Publishers, Boca Raton, FL, 1995, 13.

La Point, T.W. and Fairchild, J.F., Evaluation of sediment contamination toxicity: the use of freshwater community structure, in *Sediment Toxicity Assessment*, Burton, G.A., Jr., Ed., Lewis Publishers, Boca Raton, FL, 1992, 67.

Lee, H., II, Models, muddles and mud: predicting bioaccumulation of sediment-associated pollutants, in *Sediment Toxicity Assessment*, Burton, G.A., Jr., Ed., Lewis Publishers, Boca Raton, FL, 1992, 267.

Lenat, D.R., A biotic index for the southeastern U.S.: derivation and list of tolerance values, with criteria for assigning water-quality ratings, *J. North Am. Benthol. Soc.*, 12, 279, 1993.

Lockhart, W.L. and Metner, D.A., Fish serum chemistry as a pathological tool, in *Contaminant Effects on Fishes*, Vol. 16, Cairns, V.W., Hodson, P.V., and Nriagu, J.D., Eds., John Wiley & Sons, New York, 1984, 73.

Long, E.R. and Chapman, P.M., A sediment quality triad: measures of sediment contamination, toxicity and infaunal community composition in Puget Sound, *Mar. Pollut. Bull.*, 16, 405, 1985.

Ludke, J.L., Hill, E.F., and Dieter, M.P., Cholinesterase (ChE) response and related mortality among birds fed ChE inhibitors, *Arch. Environ. Contam. Toxicol.*, 3, 1, 1958.

Mac, M.J. and Schmitt, C.J., Sediment bioaccumulation testing in fish, in *Sediment Toxicity Assessment*, Burton, G.A., Jr., Eds., Lewis Publishers, Boca Raton, FL, 1992, 67.

Maine Department of Environmental Protection, Methods for Biological Sampling and Analysis of Maine's Waters, Maine Department of Environmental Protection, Augusta, 1987.

Maki, A.W., Stewart, K.W., and Silvey, J.K.G., The effects of Dibrom on respiratory activity of the stonefly, *Hydroperla crosbyi*, hellgrammite, *Corydalus cornutus* and the golden shiner, *Notemigonus crysileucas*, *Trans. Am. Fish. Soc.*, 102, 806, 1973.

Marchant, R., A subsampler for samples of benthic invertebrates, *Bull. Aust. Soc. Limnol.*, 12, 49, 1989.

Margalef, R., Diversidad et especies en las comunidades naturales, *Publ. Inst. Biol. Apl. Barcelona*, 6, 59, 1951.

Mason, W.T., Jr., A rapid procedure for assessment of surface mining impacts to aquatic life, in *Coal Conference and Expo V, Proceedings of a Symposium*, McGraw-Hill, New York, 1979, 310.

Mayer, F.L., Versteeg, D.J., McKee, M.J., Folmar, L.C., Graney, R.L., McCume, D.C., and Rattner, B.A., Physiological and nonspecific biomarkers, in *Biomarkers: Biochemical, Physiological and Histopathological Markers of Anthropogenic Stress*, Hugget, R.J., Kimerle, R.A., Mehrle, P.M., and Bergmen, H.L., Eds., Lewis Publishers, Boca Raton, FL, 1992, 5.

McCarthy, J.F. and Shugart, L.R., *Biomarkers of Environmental Contamination*, Lewis Publishers, Boca Raton, FL, 1990, 457.

McMahon, G., Huber, L.J., Moore, M.J., Stegeman, J.J., and Wogan, G.N., c-K-*ras* oncogenes: prevalence in livers of winter flounder from Boston Harbor, in *Biological Markers of Environmental Contaminants*, McCarthy, J.F. and Shugart, L.R., Eds., Lewis Publishers, Boca Raton, FL, 1990, 229.

Medine, A.J., Porcella, D.B., and Adams, V.D., Heavy metal and nutrient effects on sediment oxygen demand in three-phase aquatic microcosms, in *Microcosms in Ecological Research*, Giesy, J.P.J., Ed., CONF-781101, National Technical Information Service, Springfield, VA, 1980, 279.

Melancon, M.J., Bioindicators used in aquatic and terrestrial monitoring, in *Handbook of Ecotoxicology*, Hoffman, D.J., Rattner, B.A., Burton, G.A., Jr., and Cairns, J., Jr., Eds., Lewis Publishers, Boca Raton, FL, 1995, 220.

Melancon, M.J., Alscher, R., Benson, W., Kruzynski, G., Lee, R.F., Sikka, H.C., and Spies, R.B., Metabolic products as biomarkers, in *Biomarkers: Biochemical, Physiological and Histopathological Markers of Anthropogenic Stress,* Hugget, R.J., Kimerle, R.A., Mehrle, P.M., and Bergmen, H.L., Eds., Lewis Publishers, Boca Raton, FL, 1992, 87.

Menhinick, E.F., A comparison of some species–individuals diversity indices applied to samples of field insects, *Ecology*, 45, 859, 1964.

Meyers-Schone, L., Shugart, L.R., Beauchamp, J.J., and Walton, B.T., Comparison of two freshwater turtle species as monitors of radionuclide and chemical contamination: DNA damage and residue analysis, *Environ. Toxicol. Chem.*, 12, 1487, 1993.

Milbrink, G., Characteristic deformities in tubificid oligochaetes inhabiting polluted bays of Lake Vanern, southern Sweden, *Hydrobiologia*, 106, 169, 1983.

Millington, P.J. and Walker, K.F., Australian freshwater mussel *Velesunio ambiguus* (Phillipi) as a biological monitor for zinc, iron and manganese, *Aust. J. Mar. Freshwater Res.*, 34, 873, 1983.

Minchin, P.R., An evaluation of the relative robustness of techniques for ecological ordination, *Vegetatio*, 69, 89, 1987.

Moss, D., Furse, M.T., Wright, J.F., and Armitage, P.D., The prediction of the macroinvertebrate fauna of unpolluted running-water sites in Great Britain using environmental data, *Freshwater Biol.*, 17, 41, 1987.

Munawar, M. and Munawar, I.F., Phytoplankton bioassays for evaluation toxicity of *in situ* sediment contaminants, *Hydrobiologia*, 149, 87, 1987.

Muncaster, B.W., Hebert, P.D.N., and Lazar, R., Biological and physical factors affecting the body burden of organic contaminants in freshwater mussels, *Arch. Environ. Contam. Toxicol.*, 19, 25, 1990.

Murty, A.S. and Devi, A.P., The effect of endosulfan and its isomers on tissue protein, glycogen and lipids in the fish *Channa punctatus*, *Pestic. Biochem. Physiol.*, 17, 280, 1982.

Naylor, C., Maltby, L., and Calow, P., Scope for growth in *Gammarus pulex*, a freshwater benthic detritivore, *Hydrobiologia*, 188/189, 517, 1989.

Nebeker, A.V., Cairns, M.A., Gakstatter, J.H., Malueg, K.W., Schuytema, G.S., and Krawczyk, D.F., Biological methods for determining toxicity of contaminated freshwater sediments to invertebrates, *Environ. Toxicol. Chem.*, 3, 617, 1984.

Needham, P.R. and Usinger, R.L., Variability in the macrofauna of a single riffle in Prosser Creek, California, as indicated by the Surber sampler, *Hilgardia*, 24, 383, 1956.

Norris, R.H., Rapid biological assessment, selecting reference sites and natural variability, in Proc. Joint South African/Australian Workshop: Classification of Rivers and Environmental Health Indicators, Capetown, 1994, 129.

Norris, R.H., Biological monitoring: the dilemma of data analysis, *J. North Am. Benthol. Soc.*, 14, 440, 1995.

Norris, R.H. and Georges, A., Analysis and interpretation of benthic macroinvertebrate surveys, in *Freshwater Biomonitoring and Benthic Macroinvertebrates*, Rosenberg, D.M. and Resh, V.H., Eds., Chapman and Hall, New York, 1993, 234.

Ohio Environmental Protection Agency, Biological Criteria for the Protection of Aquatic Life, Vol. 1, The Role of Biological Data in Water Quality Assessment, Ohio EPA, Columbus, 1987.

Ormerod, S.J. and Edwards, R.W., The ordination and classification of macroinvertebrate assemblages in the catchment of the River Wye in relation to environmental factors, *Freshwater Biol.*, 17, 533, 1987.

Painter, S., Regional Variability in Sediment Background Metal Concentrations and the Ontario Sediment Quality Guidelines, NWRI Report No. 92-85, Environment Canada, Burlington, Ontario, 1992.

Pasteris, A., Bonacina, C., and Bonomi, G., Age, stage and size structure as population state variables for *Tubifex tubifex* (Oligochaeta: Tubificidae), *Hydrobiologia*, 334, 125, 1996.

Peakall, D.B., *Animal Biomarkers as Pollution Indicators*, Chapman and Hall, London, 1992, 291.

Pedersen, B.V., The effect of anoxia on the survival of chromosomal variants in the larvae of the midge *Chironomus plumosus* L. (Diptera: Chironomidae), *Hereditas*, 101, 75, 1984.

Persaud, D., Jaagumagi, R., and Hayton, A., Guidelines for the Protection and Management of Aquatic Sediment Quality in Ontario, Water Resources Branch, Ontario Ministry of Environment and Energy, Toronto, Ontario, 1992.

Petering, D.H., Goodrich, M., Hodgman, W., Krezoski, S., Weber, D., Shaw, C.F., III, Spieler, R., and Zettergren, L., Metal binding proteins and peptides for the detection of heavy metals in aquatic organisms, in *Biomarkers of Environmental Contamination*, Shugart, L.R. and McCarthy, J.F., Eds., Lewis Publishers, Boca Raton, FL, 1990, 239.

Petersen, L.B.M. and Petersen, R.C., Jr., Effect of kraft pulp mill effluent and 4,5,6 trichloroguaiacol on the net spinning behaviour of *Hydropsyche angustipennis* (Trichoptera), *Ecol. Bull.*, 36, 68, 1984.

Phillips, D.J.H., *Quantitative Aquatic Biological Indicators: Their Use to Monitor Trace Metal and Organochlorine Pollution*, Applied Science, London, 1980, 488.

Pierce, K.V., McCain, B.B., and Willings, S.R., Pathology of hepatoma and other liver abnormalities in English sole (*Parophrys vetulis*) from the Duwamish River estuary, Seattle, Washington, *J. Natl. Cancer Inst.*, 50, 1445, 1978.

Pinkham, C.F.A. and Pearson, J.G., Applications of a new coefficient of similarity to pollution surveys, *J. Water Pollut. Control Fed.*, 48, 717, 1976.

Plafkin, J.L., Barbour, M.T., Porter, K.D., Gross, S.K., and Hughes, R.M., Rapid Bioassessment Protocols for Use in Streams and Rivers. Benthic Macroinvertebrates and Fish, Office of Water Regulations and Standards, U.S. Environmental Protection Agency, Washington, D.C., 1989.

Power, E. and Chapman, P., Assessing sediment quality, in *Sediment Toxicity Assessment*, Burton, G.A., Jr., Ed., Lewis Publishers, Boca Raton, FL, 1992, 1.

Prichard, P.M. and Bourquin, A.W., Microbial toxicity studies, in *Fundamentals of Aquatic Toxicology*, Rand, G.M. and Petrocelli, S.R., Eds., Hemisphere, New York, 1985, 177.

Randerath, K., Reddy, M., and Gupta, R.C., ^{32}P-postlabelling analysis for DNA damage, *Proc. Natl. Acad. Sci. U.S.A.*, 78, 6126, 1981.

Rattner, B.A. and Fairbrother, A., Biological diversity and the influence of stress on cholinesterase activity, in *Cholinesterase Inhibiting Insecticides: Their Impact on Wildlife and the Environment*, Mineau, P., Ed., Elsevier Science, Amsterdam, 1991, 348.

Reddy, V.R., Christenson, W.R., and Piper, W.N., Extraction and isolation by high performance liquid chromatography of uroporphyrin and corproporphyrin isomers from biological tissue, *J. Pharmacol. Method.*, 17, 51, 1987.

Reice, S.R. and Wohlenberg, M., Monitoring freshwater benthic macroinvertebrates and benthic processes: measures for assessment of ecosystem health, in *Freshwater Biomonitoring and Benthic Macroinvertebrates*, Rosenberg, D.M. and Resh, V.H., Eds., Chapman and Hall, New York, 1993, 287.

Resh, V.H., Sampling variability and life history features: basic considerations in the design of aquatic insect studies, *J. Fish. Res. Board Can.*, 36, 290, 1979.

Resh, V.H., Freshwater benthic macroinvertebrates and rapid assessment procedures for water quality monitoring in developing and newly industrialized countries, in *Biological Assessment and Criteria: Tools for Water Resource Planning and Decision Making*, Davis, W.S. and Simon, T., Eds., Lewis Publishers, Boca Raton, FL, 1995, 167.

Resh, V.H. and McElravy, E.P., Contemporary quantitative approaches to biomonitoring using benthic macroinvertebrates, in *Freshwater Biomonitoring and Benthic Macroinvertebrates*, Rosenberg, D.M. and Resh, V.H., Eds., Chapman and Hall, New York, 1993, 159.

Resh, V.H. and Unzicker, J.D., Water quality monitoring and aquatic organisms: the importance of species identification, *J. Water Pollut. Control Fed.*, 47, 9, 1975.

Resh, V.H., Rosenberg, D.M., and Feminella, J.W., The processing of benthic samples: responses to the 1983 NABS questionnaire, *Bull. North Am. Benthol. Soc.*, 2, 5, 1985.

Resh, V.H., Norris, R.H., and Barbour, M.T., Design and implementation of rapid assessment approaches for water resource monitoring using benthic macroinvertebrates, *Aust. J. Ecol.*, 20, 108, 1995.

Reynoldson, T.B. and Zarull, M.A., An approach to the development of biological sediment guidelines, in *Ecological Integrity and Management of Ecosystems*, Francis, G., Kay, J., and Woodley, S., Eds., St. Lucie Press, Boca Raton, FL, 1993, 177.

Reynoldson, T.B., Bailey, R.C., Dayand, K.E., and Norris, R.H., Biological guidelines for freshwater sediment based on Benthic Assessment of Sediment (the BEAST) using a multivariate approach for predicting biological state, *Aust. J. Ecol.*, 20, 198, 1995.

Rosenberg, D.M. and Resh, V.H., *Freshwater Biomonitoring and Benthic Macroinvertebrates*, Chapman and Hall, New York, 1993, 488.

Ross, L.T. and Jones, D.A., Eds., Biological Aspects of Water Quality in Florida, Technical Series Vol. 4, No. 3, Department of Environmental Regulation, State of Florida, Tallahassee, 1979.

Ross, P., Jarry, V., and Sloterdijk, H., A rapid bioassay using the green alga *Selenastrum capricornutum* to screen for toxicity in St. Lawrence River sediment elutriates, in *Functional Testing of Aquatic Biota for Estimating Hazards of Chemicals, STP988*, Cairns, J.J., Jr. and Pratt, J.R., Eds., American Society for Testing and Materials, Philadelphia, 1988, 68.

Russell, R.W. and Gobas, F.A.P.C., Calibration of the freshwater mussel, *Elliptio complanata,* for quantitative biomonitoring of hexachlorobenzene and octachlorostyrene in aquatic systems, *Bull. Environ. Contam. Toxicol.*, 43, 576, 1989.

Sanders, B.M., Stress proteins: potential as multitiered biomarkers, in *Biomarkers of Environmental Contamination*, Shugart, L.R. and McCarthy, J.F., Eds., Lewis Publishers, Boca Raton, FL, 1990, 165.

Sanders, B.M., Stress proteins in aquatic organisms: an environmental perspective, *Crit. Rev. Toxicol.*, 23, 49, 1993.

Sasson-Brickson, G. and Burton, G.A., Jr., *In situ* and laboratory sediment toxicity testing with *Ceriodaphnia dubia*, *Environ. Toxicol. Chem.*, 10, 201, 1991.

Sayler, G.S., Sherrill, T.W., Perkins, R.E., Mallory, L.M., Shiaris, M.P., and Pedersen, D., Impact of coal-coking effluent on sediment microbial communities: a multivariate approach, *Appl. Environ. Microbiol.*, 44, 1118, 1979.

Scanferlato, V.S. and Cairns, J.J., Jr., Effect of sediment-associated copper on ecological structure and function of aquatic microcosms, *Aquat. Toxicol.*, 18, 23, 1990.

Sebastien, R., Rosenberg, D.M., and Wiens, A.P., A method for subsampling unsorted benthic macroinvertebrates by weight, *Hydrobiologia*, 157, 69, 1988.

Shackleford, B., Rapid Bioassessments of Lotic Macroinvertebrate Communities, Biocriteria Development, Biomonitoring Section, Arkansas Department of Pollution Control and Ecology, Little Rock, 1988.

Shannon, C.E., A mathematical theory of communication, *Bell System Tech. J.*, 27, 379, 1948.

Sharma, R.P., *Immunologic Considerations in Toxicology*, Vol. I and II, CRC Press, Boca Raton, FL, 1981.

Shugart, L.R., Quantitation of chemically induced damage to DNA of aquatic organisms by alkaline unwinding assay, *Aquat. Toxicol.*, 13, 43, 1988.

Shugart, L.R., 5-Methyl deoxycytidine content of DNA from bluegill sunfish (*Lepomis macrochirus*) exposed to benzo[a]pyrene, *Environ. Toxicol. Chem.*, 9, 205, 1990.

Shugart, L.R., Bickham, J., Jackim, G., McMahon, G., Ridley, W., Stein, J., and Steinert, S., DNA alterations, in *Biomarkers: Biochemical, Physiological and Histopathological Markers of Anthropogenic Stress*, Hugget, R.J., Kimerle, R.A., Mehrle, P.M., and Bergmen, H.L., Eds., Lewis Publishers, Boca Raton, FL, 1992, 125.

Simpson, E.H., Measurement of diversity, *Nature*, 136, 688, 1949.

Southerland, M.T. and Stribling, J.B., Status of biological criteria development and implementation, in *Biological Assessment and Criteria: Tools for Water Resource Planning and Decision Making*, Davis, W.S. and Simon, T.P., Eds., Lewis Publishers, Boca Raton, FL, 1995, 81.

Stegeman, J.J., Teng, F.Y., and Snowberger, E.A., Induced cytochrome P450 in winter flounder (*Pseudopleuronectes americanus*) from coastal Massachusetts evaluated by catalytic assay and monoclonal antibody probes, *Can. J. Fish. Aquat. Sci.*, 44, 1270, 1987.

Stegeman, J.J., Brouwer, M., Di Giulio, R.T., Forlin, L., Fowler, B.A., Sanders, B.M., and Van Veld, P.A., Enzyme and protein synthesis as indicators of contaminant exposure, in *Biomarkers: Biochemical, Physiological and Histopathological Markers of Anthropogenic Stress*, Hugget, R.J., Kimerle, R.A., Mehrle, P.M., and Bergmen, H.L., Eds., Lewis Publishers, Boca Raton, FL, 1992.

Stewart-Oaten, A., Murdoch, W., and Parker, K., Environmental impact assessment: "pseudoreplication" in time? *Ecology*, 67, 929, 1986.

Stewart-Oaten, A., Bence, J.R., and Osenberg, C.W., Assessing effects of unreplicated perturbations: no simple solutions, *Ecology*, 73, 1396, 1992.

Sutcliffe, W.H.J., Relationship between growth rate and ribonucleic acid concentration in some invertebrates, *J. Fish. Res. Board Can.*, 27, 606, 1970.

Thomas, P., Effect of cadmium exposure on plasma cortisol levels and carbohydrate metabolism in mullet (*Mugil cephalus*), *J. Endocrinol.*, 94 (Suppl.), 35, 1982.

Van Contfort, J., De Graeve, J., and Gielen, J.E., Radioactive assay for aryl hydrocarbon hydroxylase: improved method and biological importance, *Biochem. Biophys. Res. Commun.*, 79, 505, 1977.

Varanasi, U., Reichert, W.L., Eberhart, B.T., and Stein, J., Formation and persistence of benzo[a]pyrene-diolepoxide-DNA adducts in liver of English sole (*Parophrys vetulus*), *Chem. Biol. Interact.*, 69, 203, 1989.

Verma, S.R., Rani, S., Tonk, I.P., and Dalela, R.C., Pesticide-induced dysfunction in carbohydrate metabolism in three freshwater fish, *Environ. Res.*, 32, 127, 1983.

Warriner, J.E., Matthews, E.S., and Weeks, B.A., Preliminary investigations of the chemiluminescent response in normal and pollutant-exposed fish, *Mar. Environ. Res.*, 24, 281, 1988.

Warwick, R.M., The level of taxonomic discrimination required to detect pollution effects on marine benthic communities, *Mar. Pollut. Bull.*, 19, 259, 1988.

Warwick, R.M., Environmental impact studies on marine communities: pragmatical considerations, *Aust. J. Ecol.*, 18, 63, 1993.

Warwick, R.M. and Clarke, K.R., Comparing the severity of disturbance: a meta-analysis of marine macrobenthic community data, *Mar. Ecol. Prog. Ser.*, 92, 221, 1993.

Warwick, W.F., Morphological abnormalities in Chironomidae (Diptera) larvae as measures of toxic stress in freshwater ecosystems: indexing antennal deformities in *Chironomus meigen*, *Can. J. Fish Aquat. Sci.*, 42, 1881, 1985.

Warwick, W.F., Morphological deformities in Chironomidae (Diptera) larvae as biological indicators of toxic stress, in *Toxic Contaminants and Ecosystem Health: A Great Lakes Focus*, Evans, M.S., Ed., John Wiley, New York, 1988, 281.

Warwick, W.F., Indexing deformities in ligulae and antennae of *Procladius* larvae (Diptera: Chironomidae): application to contaminant stressed environments, *Can. J. Fish. Aquat. Sci.*, 48, 1151, 1991.

Washington, H.G., Diversity, biotic and similarity indices: a review with special relevance to aquatic ecosystems, *Water Res.*, 18, 653, 1984.

Watson, C.F., Baer, K.N., and Benson, W.H., Dorsal gill incision: a simple method for obtaining blood samples in small fish, *Environ. Toxicol. Chem.*, 8, 457, 1989.

Weeks, B.A., Anderson, B.P., Goven, A.J., Dufour, A.P., Fairbrother, A., Lahuis, G.P, and Peters, G., Immunological biomarkers to assess environmental stress, in *Biomarkers: Biochemical, Physiological, and Histopathological Markers of Anthropogenic Stress*, Hugget, R.J., Kimerle, R.A., Mehrle, P.M., and Bergmen, H.L., Eds., Lewis Publishers, Boca Raton, FL, 1992, 211.

Weiss, C.M., The determination of cholinesterase in the brain tissue of three species of fresh water fish and its inactivation *in vivo*, *Ecology*, 39, 194, 1958.

Weiss, C.M., Physiological effect of organic phosphorus insecticides on several specie of fish, *Trans. Am. Fish. Soc.*, 90, 143, 1961.

Wiederholm, T., Wiederholm, A.M., and Milbrink, G., Bulk sediment bioassays with five species of fresh water oligochaetes, *Water Air Soil Pollut.*, 36, 131, 1987.

Winget, R.N. and Mangum, F.A., Biotic Condition Index: Integrated Biological, Physical and Chemical Stream Parameters for Management, U.S. Forest Service Intermountain Region, U.S. Department of Agriculture, Ogden, UT, 1979.

Wong, S., Fournier, M., Corderre, D., Banska, W., and Krzystyniak, K., Environmental immunotoxicology, in *Animal Biomarkers as Pollution Indicators,* Peakall, D., Ed., Chapman and Hall, London, 1992, 167.

Wright, J.F., Moss, D., Armitage, P.D., and Furse, M.T., A preliminary classification of running-water sites in Great Britain based on macroinvertebrate species and the prediction of community type using environmental data, *Freshwater Biol.,* 14, 221, 1984.

Wright, J.F., Armitage, P.D., Furse, M.T., and Moss, D., A new approach to the biological surveillance of river quality using macroinvertebrates, *Int. Ver. Theor. Angew. Limnol. Verh.,* 23, 1548, 1988.

Wrona, F.J., Culp, J.M., and Davies, R.W., Macroinvertebrate subsampling, a simplified apparatus and approach, *Can. J. Fish. Aquat. Sci.,* 39, 1051, 1982.

Yoder, C.O. and Rankin, E.T., Biological criteria development program and implementation in Ohio, in *Biological Assessment and Criteria: Tools for Water Resource Planning and Decision Making,* Davis, W.S. and Simon, T.P., Lewis Publishers, Boca Raton, FL, 1995, 109.

Zarull, M.A. and Reynoldson, T.B., A management strategy for contaminated sediment assessment and remediation, *Water Pollut. Res. J. Can.,* 27, 871, 1992.

Zelinka, M. and Marvan, P., Zur Präzisierung der biologischen Klassifikation der reinheit fliessender Gewässer, *Arch. Hydrobiol.,* 57, 389, 1961.

chapter five

Sediment certified reference materials

Venghuot F. Cheam

5.1 Introduction

It is often said that bad data are worse than no data at all. After all, bad data lead to bad conclusions and decision making. "Even under very conservative estimates, costs resulting from unreliable measurements run into billions of dollars per year" (Cali, 1976). On the other hand, it can also be said that data are as good as the quality assurance associated with the data generation process. It is not sufficient to merely generate data. The data must be ensured as reliable and interpreted with wisdom and understanding.

Although the funds allocated for environmental studies have steadily increased over the past two to three decades, more dramatic is the increase in the fraction of those funds dedicated to quality assurance/quality control (QA/QC) programs, which have been designed to pursue data reliability. For instance, funds for the development of clean rooms, certified reference materials (CRMs), more sensitive instruments, and more reliable analytical methods are just a few examples of resources allocated for QA/QC.

An effective QA program, whose primary aim is to ensure that the generated analytical data are a true, authentic representation of the sample content just before the sampling takes place, comprises several domains. Some key ones are the following (Uriano and Gravatt, 1977; Hunter, 1980; Lawrence et al., 1982; Chau, 1983; Taylor, 1985):

1. Sample integrity
2. Analytical methods
3. Well-trained and experienced analysts
4. Intralaboratory QC program
5. Interlaboratory QC program

1-56670-343-3/99/$0.00+$.50

Sample integrity is the first and foremost domain that must be assured, for if it falters, the rest of QA is immaterial or, at best, relative. Sample integrity is achieved by using proven procedures/protocols for labware cleaning, type of containers, sampling procedure, procedure and field blanks, preservation, clean room practice, and properly trained personnel. Each procedure is specific to a class of analytes of interest. For example, containers for samples collected for the determination of trace metals require a specific, lengthy acid-cleaning procedure (Tramontano et al., 1987), which is different from the cleaning procedure for containers for samples collected for the determination of organic contaminants. Similarly, containers for samples collected for the determination of trace metals should generally be plastic, whereas those for samples for the determination of organic contaminants should be made of glass. If a certain procedure becomes dubious with time, an improved or new one must be developed and ascertained.

Analytical methods must be properly developed and validated against pertinent CRMs and reference methods. The proper methods and equipment in sound working condition are the right tools for an effective QA program.

Well-trained and experienced analysts are another prerequisite, as they have acquired the expertise and the maturity to effectively apply the various facets of interweaving procedures. It has often been observed via interlaboratory QC studies that laboratories which consistently perform well with experienced analysts would perform less well with freshly trained analysts.

Intralaboratory QC is a fundamental routine that a laboratory must establish and follow to assure in-house operations are in control. Key QC activities incorporate proper/acceptable blank, duplicate run, spike recovery, control charts, uniformly spaced standard recalibration, confirmation by another validated method, and use of relevant CRMs. An intralaboratory QC program without pertinent CRMs, no matter how rigorous it may be, may provide in-house (i.e., within-laboratory) precision but not necessarily accuracy and thus not necessarily data reliability. Therefore, the QC program should also include development of pertinent in-house/secondary reference materials (even certified ones), if such CRMs are not available commercially.

Interlaboratory QC must also be an integral part of an effective QA program, for even if the above aforementioned four domains are in check, the data are generated from only one laboratory, one method, and possibly one analyst. No laboratory can exist in isolation without outside interaction. It must know its performance with respect to its counterparts. This involves participating regularly in one or more multilaboratory intercomparison studies to assess the effectiveness of its in-house QC and to obtain a measure of performance among peers (Cheam et al., 1988; Quevauviller et al., 1994a,b; others discussed in more detail below). Again, an interlaboratory study may provide interlaboratory (between-laboratory) comparability but not necessarily accuracy and thus not necessarily data reliability if pertinent CRMs are not incorporated in the study or used somewhere in the process. Intercomparison studies are usually run by a qualified "QA group," which has several important activities, one of which is the development of pertinent CRMs (Lawrence et al., 1982; Community Bureau of Reference, 1994; Analytical Quality Control Services, 1995).

From the above, it is obvious that CRMs are a key requirement in most of the above domains. Many authors have dealt with the subject of reference materials, for example, Cali et al. (1975), Taylor (1985), Roelandts (1989), Cantillo (1992), Govindaraju (1994), and Willie and Berman (1994). This chapter deals specifically with sediment CRMs in more detail. Sediment CRMs are relatively young compared to other types of CRMs. The first sediment reference material was issued only some 70 years after the first, mostly industry related, CRMs were on the market at the turn of the century, when environmental issues were not as important as they are at present. The preparation, certification, literature review, and availability of most, if not all, sediment CRMs listed in the scientific literature are presented in this chapter. Readily useful information on each CRM is given to assist analysts in identifying the appropriate ones for their applications.

5.2 Certified reference materials: literature review

The National Institute of Standards and Technology (NIST, formerly the National Bureau of Standards) issued the first four NIST Standard Reference Materials (SRMs) for cast iron in chip form in 1906, one of which (SRM 4M) has been reissued numerous times and is still in use to provide accuracy for the determination of gray cast irons (Cali, 1976; Uriano, 1980). NIST has been the largest producer of CRMs in the world. (Note that the term SRM refers to a CRM produced and issued by NIST). By 1911, approximately 140 SRMs had been produced for purity or compositional determination of steels, ores, brasses, and sugars or for combustion applications. By 1951, NIST had about 500 different CRMs available. In 1979, some 38,000 SRMs were distributed. In England, a number of metal and ore reference materials had been issued as "British Chemical Standards" by 1920 (Uriano, 1980).

Globally, there has been a rapid increase in the development of CRMs in recent years, particularly in developed countries. Different types of CRMs exist for physical properties such as surface area, color, wavelength, etc.; for compositional properties, such as organic and inorganic substances in various media; and for engineering properties, such as sieve sizing, color standards, etc. CRMs also exist in all three states: gaseous (e.g., N_2, Ar, etc.), liquid (e.g., sea water, fresh water, etc.), and solid (e.g., tissues, rocks, soils, sediments, etc.) (Uriano and Gravatt, 1977, 1980; Taylor, 1985; Cantillo, 1992;

Willie and Berman, 1994). The first sediment CRM may have been NIST-SRM 4350, a river sediment certified for 12 radionuclides which was issued around the mid-1970s (Cali, 1976). The development of sediment CRMs intensified with an increase in research studies on aquatic sediments, whether of a geochemical nature, interactions at the sediment–water interface or interstitial waters, or sediment dredging/disposal regulatory tests. Diversified sediment studies require diversified sediment matrices, analytes, and concentration levels. The more diversified CRMs are, the more one of them is closer to the ideal state of matrix and concentration matching with samples. This should maximize the accuracy of measured analytical data.

Manheim et al. (1976) of the U.S. Geological Survey (USGS) reported a marine sediment standard, MAG-1, a gray-brown very fine-grained clayey mud collected from the Wilkinson Basin of the Gulf of Maine. It was the first sediment standard of its kind among the many rock standards reported since 1951 (Fairbairn et al., 1951). Note that the hot spring deposit sample, GXR-3 of the USGS, reported by Alcott and Lakin (1975) could be considered and included as a sediment reference material. However, there appears to be some significant variation between bottles, and much caution should be exercised when using it (Govindaraju, 1994).

NIST issued another certified river sediment reference material, SRM 1645, dated November 1978 (National Bureau of Standards [NBS], 1978). The SRM was prepared from dredged bottom sediment of the Indiana Harbor Canal, Chicago, Illinois. Other sediment CRMs are SRM 1646 and SRM 2704 for inorganic constituents; SRM 1941 and SRM 1939 for polyaromatic hydrocarbons (PAHs) and polychlorinated biphenyls (PCBs); and SRM 4350, SRM 4350B, and SRM 4354 for radionuclides.

For over 30 years, the International Atomic Energy Agency (IAEA) in Austria, through Analytical Quality Control Services (AQCS), has initiated and supported improvements in the accuracy of analytical chemistry and radiometric measurements and their traceability to basic standards. They carried out an intercomparison run in 1977 on a candidate CRM sediment sample, SL-1. The sediment was collected at the Sardis Reservoir, Panola County, Mississippi (IAEA, 1977; Dybczynski and Suschny, 1979). SL-1 was certified and distributed in 1981 (IAEA, 1980; AQCS, 1992). Other sediment CRMs include SL-3, IAEA-356, and SD-M-2/TM for inorganic constituents; IAEA-356 for methyl mercury and IAEA-357 for PAHs, PCBs, and other organics; and SL-2, SD-N-2, IAEA-135, IAEA-300, IAEA-313, IAEA-314, IAEA-315, IAEA-367, and IAEA-368 for various radionuclides.

The Canadian Certified Reference Materials Project (CCRMP), administered by the Canada Centre for Mineral and Energy Technology (CANMET), Ottawa, Ontario, had its beginning in 1955; by 1962, it had released for sale three phosphor bronze discs (Steger, 1984). Several rocks, soils, and slag reference materials had also been issued (Steger, 1980). In about 1983, eight sediment samples were proposed in collaboration with the Geological Survey of Canada (Abbey, 1983): four of lake origin (LKSD-1 to LKSD-4) and four of stream origin (STSD-1 to STSD-4). Lynch (1990) and Bowman (1990) published provisional values for both total and extractable metal constituents for all eight sediment samples.

The National Water Research Institute (NWRI), Burlington, Ontario, initiated its development of reference standards for waters and sediments in about 1975, and by 1979 the first paper of a series on the preparation and purification of photomirex was published (Chau and Thompson, 1979). Preparation of the first sediment reference materials, WQB-1 and WQB-2, was initiated in 1974 (NWRI, 1988), and its certification was finished in 1983 (Cheam and Chau, 1984; NWRI, 1990a). Other sediment CRMs for both organic and inorganic parameters followed (NWRI, 1990b,c, 1992a,b,c, 1995b): EC-1, EC-2, and EC-3 for PAHs, PCBs, and chlorobenzenes; DX-1 and DX-2 for polychlorinated dibenzo-*p*-dioxins (PCDDs) and polychlorinated dibenzofurans (PCDFs); WQB-3 for trace metals; and several other uncertified sediment reference materials.

Since 1981, the National Research Council of Canada (NRCC), Ottawa, Ontario, has issued more than 30 matrix standards and CRMs for shellfish toxins, PCBs, PAHs, PCDDs, PCDFs, organometallics, and trace metals (Berman et al., 1994). The first sediment CRMs released were MESS-1 and BCSS-1, which were announced in the *Geostandards Newsletter* in 1981 by Berman. Preparation of these two materials began in 1978 (Guevremont and Jamieson, 1982). Other sediment CRMs are MESS-2, BEST-1, and PACS-1 for inorganic parameters (PACS-1 is also for organotin); CS-1, HS-1, and HS-2 for PCBs and PCB congeners; and several uncertified sediment reference materials.

The Community Bureau of Reference (BCR, presently Measurements and Testing Programme) of the European Commission has produced various CRMs over the last 15 years (Quevauviller, 1994). These materials are of environmental matrix, food and agriculture, biomedical, physical properties, and industrial raw materials (BCR, 1994 and 1995 addendum). The coastal sediment CRM 462 was successfully certified via a recent interlaboratory study for di- and tributyltins (Quevauviller et al., 1994a). Other sediment materials include CRM 277 (estuarine sediment), CRM 280 (lake sediment), and CRM 320 (river sediment).

The National Institute for Environmental Studies (NIES) of the Japan Environmental Agency has produced a number of CRMs for the past several years, such as chlorella, vehicle exhaust particulates, rice flour and fish tissue, and a pond sediment called NIES No. 2 (NIES, 1981; Iwata et al., 1983a,b; Okamoto and Fuwa, 1985). The sediment was collected from the surface of the bottom of Sanshiro pond within the grounds of Tokyo University in 1977 and was certified in July 1981.

The South African Committee for CRMs (SACCRM) was appointed in 1974 to direct the production of reference materials and has produced several rock standards (SACCRM, 1992; Ring, 1993; Frick, 1981). In addition to these, three stream sediment CRMs (i.e., SARM-46, SARM-51, and SARM-

52) were prepared in 1978 and were apparently certified in 1992 (Ring, 1993).

The Institute of Geophysical and Geochemical Exploration (IGGE) and the Institute of Rocks and Mineral Analysis (IRMA) of the People's Republic of China prepare and certify reference materials. The first eight stream sediment samples, GSD 1 to 8, were prepared in 1980 (Mingcai et al., 1980) and certified in 1983 using interlaboratory data (Xie et al., 1985a). Another suite of four stream sediments, GSD 9 to 12, as well as eight soil samples and six rock samples were collected and processed between 1980 and 1983 (Xie et al., 1985b) and certified in 1989 (Xie et al., 1989).

The Geological Survey of Japan issued 17 rock reference samples, igneous rock series, in 1986 and 1989 (Ando et al., 1987, 1989). Another suite of rock reference samples, sedimentary rocks series, started in 1986 and was issued in 1989. The latter series included four sediment samples: a lake sediment (Jlk-1) and three stream sediments (Jsd-1, Jsd-2, Jsd-3) (Terashima et al., 1990).

In 1975, the Research Institute of Applied Physics (RIAP), Irkutsk, Russia, and the Institute of Oceanography of the Russian Academy of Sciences jointly began investigation of three marine sediment reference samples, SDO-1, SDO-2, and SDO-3. Berkovits and Lukashin (1984) reported the certification of the three samples, certified for major, minor, and trace elements. Berkovits et al. (1991) reported a series of 19 certified reference samples for clay, nodule, crust, etc. and four sediments, coded OOKO201–04, using the data from a collaborative study involving more than 100 laboratories. Preparation of the latter sediment samples, which are called anomalous and background sediments, was started in 1983 jointly by RIAP and the Institute of Geochemistry (Siberian Branch).

5.3 *Developmental procedures for certified reference materials*

5.3.1 *Criteria for selecting a certified reference material*

In selecting a CRM for use in a QA domain, a set of practical criteria should be met (Uriano and Gravatt, 1977; International Organization for Standardization, 1981; Chau, 1983; Taylor, 1985). It is scientifically sound for a CRM to:

1. Have the certified values of analytes (properties) reliably derived
2. Be homogeneous and stable with respect to analytes of interest
3. Have a comparable matrix to that of the samples
4. Have certified analyte concentrations comparable to those in the samples
5. Specify minimum usable weight and handling instructions
6. Have authentic traceability

A certified value (or best estimate of the true value, or a recommended value) of a property should be obtained using a dependable certification procedure and provided with a statistically derived uncertainty or confidence level. Ideally, both the matrix and analyte concentrations of the CRM should be identical to those of the sample to maximize the accuracy of the measurement. In reality, however, it seems inconceivable or even unreasonable to consider that such an ideal situation ever exists; therefore, in practice, one would often be forced to aim for a close match of the CRM and sample properties. For example, in analyzing lake sediment samples, it makes sense to choose one or more lake sediment CRMs rather than marine sediments or, even worse, plant/fish CRMs. Also, if the sediment samples are contaminated, CRMs collected at contaminated sites should be used to satisfy the criteria of similar matrix and analyte concentration. Instructions such as minimum weight to be used (e.g., 0.1 or 0.5 g), storage conditions (e.g., –20 or 4°C), drying before use (e.g., 110°C for 2 hr in a desiccator over P_2O_5), or how mixing should be carried out prior to the use of the CRM are necessary to preserve the integrity of CRMs. Should new findings generate uncertainty, a CRM's authentic certificate of analysis or other publications become important documents and should be used to trace the certification process to its source. The above criteria are also useful in the planning and preparation of CRMs intended for certain specific applications (see below).

A reference material (RM) is defined here as a material that has been prepared by exactly the same procedure as a CRM (see below) and sufficiently characterized to provide a certain degree of confidence in the value(s) of one or more properties. However, none of its properties have been certified.

5.3.2 *Procedures for preparing a sediment reference material*

5.3.2.1 *Selection of suitable sampling site*

Background information on potential sites for collecting sediment should be obtained from the literature. The properties of interest, such as matrix, presence, and levels of certain analytes, should be considered and carefully evaluated. Preliminary in-house analyses should be made, if called for, to clarify the presence and concentrations of different analytes before the actual sampling takes place, to ensure beforehand that the key requirements specified in the above selection criteria are satisfactorily met.

5.3.2.2 *Sample collection*

This topic has been adequately dealt with by Mudroch and Azcue (1995).

5.3.2.3 *Drying*

Preliminary drying is essential because it removes about 50% of the water from the sediment. The wet sediment is first frozen at –20°C for a minimum of 4 days in galvanized garbage containers. The containers are then brought out of the cold room and pierced uniformly to make 50 to 100 holes about

5 mm in diameter. The sediment is allowed to thaw at room temperature for at least 3 days, during which time the water drains out, leaving the sediment in a semi-dried state. The partially dried sediment is then subdivided into lots of 20 to 25 kg, which are then freeze-dried at reduced pressure and an elevated temperature in a freeze-drying oven (Chau and Lee, 1980; Cheam and Chau, 1984).

Air-drying on filter paper for about 2 weeks has also been practiced following preliminary drying by suction to remove the sediment pore water (NIES, 1981). Quevauviller et al. (1994a) first air-dried their candidate sediment for 7 days at room temperature on a cotton sheet, followed by air-drying at 55°C for 100 hr.

5.3.2.4 Homogenization and homogeneity test on bulk sediment

The aggregated dried sediment is then ground and sieved through a series of vibrating screens of ascending mesh size, such as a 50-mesh or 300-µm sieve, followed by a 100-mesh or 150-µm sieve, a 200-mesh or 71-µm sieve, and a 325-mesh or 45-µm sieve. The sieved sediment is then combined and blended, for example, in a 600-L conical steel blender for several hours (usually 8 hr) or until the sample homogeneity is adequately proven (Epstein et al., 1989; Ring, 1993). Some examples of the final screen size used are 100 mesh for the river sediment SRM 2704 (Epstein et al., 1989); 200 mesh for the pond sediment NIES No. 2 (NIES, 1981), for a Great Lake sediment WQB-1 (Cheam and Chau, 1984), and for stream sediments GSD 1 to 8 (Xie et al., 1985a); 325 mesh for a lake sediment for PCBs analysis (Chau and Lee, 1980); and 75-µm sieve (Ring, 1993; Quevauviller et al., 1994a).

Before bottling (i.e., subsampling) can take place, homogenization of the bulk sediment must be demonstrated. The homogeneity test is usually performed by using an analysis of variance (ANOVA) for most, if not all, of the analytes of interest. Even though their analysis showed sample homogeneity with respect to sodium and silicon, Epstein et al. (1989) had to reblend their sediment because of the observed bottle-to-bottle inhomogeneity for Cr and Fe. An established homogeneity test for ore reference materials (Sutarno and Faye, 1975) was successfully used for demonstrating the homogeneity of a bulk sediment to be certified for As, Se, and Hg (Cheam and Chau, 1984). Six samples were taken from the bulk sediment in the blender: two from the top, two from the middle, and two from the bottom. Five replicate analyses were made on each sample and for each of the three elements. The two-way ANOVA technique (Snedecor and Cochran, 1967), which tests for compatibility of all between-bottle and within-bottle means, was performed on each of the three 6×5 matrices of analytical results (one matrix of results per element). The calculated variance ratios (F) for both between-bottle and within-bottle effects were found to be smaller than the critical values at a 95% confidence level, which indicates bulk sample homogeneity. Had the test shown sample inhomogeneity, the bulk sample would be reblended until satisfactory homogeneity was achieved. It is noteworthy that if the test showed inhomogeneity with respect to, for example, Hg, the certification process could still continue, but only for As and Se.

5.3.2.5 Subsampling and homogeneity test on subsamples

Subsampling of a predetermined amount of sediment, usually within the range of 10 to 100 g, is then done in precleaned glass or plastic bottles. Because bottling can take several days for a few hundred kilograms of material, frequent mixing of the bulk sediment should be carried out during bottling to ensure identical sample representation in each and every bottle. The bottled subsamples are then radiation sterilized with 2 to 2.8 Mrad of ^{60}Co to minimize biodegradation which may occur by microbiological activities (NIES, 1981; Epstein et al., 1989). Instead of radiation sterilization, Quevauviller et al. (1994a), in their certification of butyltins, recommended heat sterilization at 120°C for 2 hr following drying at 60°C for 48 hr, followed by grinding and sieving of a sediment CRM candidate. These authors and Epstein et al. (1989) subsampled the sediment after the sterilization process. Xie et al. (1989) also used heat sterilization at 120°C for 24 hr for their stream sediments certified for 72 inorganic constituents.

During bottling, a finite fraction of bottles (e.g., 1 out of every 50 bottles) is systematically put aside for testing the homogeneity of subsamples to ensure that all subsamples are the same. These samples are then analyzed for each or most of the analytes of interest. If the coefficient of variation for the analytical results of an analyte is within a set limit (e.g., ±5%) the subsamples are considered homogeneous with respect to that specific analyte. The stability study and the certification process can follow simultaneously.

5.3.2.6 Storage and stability

Sediment materials intended for the determination of organic compounds are usually stored at –20°C. For the determination of butyltins, the sediments are stored at 4°C in the dark (Quevauviller et al., 1994a). For trace metals, it is usually sufficient to store the materials in a cool room in the dark. Long-term stability of the properties of interest must be monitored and confirmed. For example, the long-term stability of PCBs, chlorobenzenes, and PAHs for two sediment CRMs stored at 4°C in the dark has been recently reported to be holding for more than 10 years (Stokker and Kaminski, 1995).

5.3.3 Procedures for certifying a sediment reference material

There are several certification procedures used by the various CRM-producing agencies.

5.3.3.1 Definitive method within a laboratory

A definitive method is defined as "the most accurate method available to measure a given chemical property" (Uriano and Gravatt, 1977). This procedure is preferred by NIST for certifying its reference materials (SRMs, which are CRMs specific to NIST). However, other producers call such materials CRMs. The method used at NIST is two or more analysts working independently to obtain the true concentration of analytes, where the uncertainty is basically the precision of the method. An example of a definitive method is isotope dilution mass spectrometry, used to determine calcium in serum (Moore and Machlan, 1972). This method was also used to certify thallium, thorium, and uranium in the river sediment reference material SRM 1645. The CCRMP also uses this approach to certify the uranium content in three ores (Steger, 1984). A variation of this procedure is the use of a definitive method by two or more laboratories.

5.3.3.2 *Two or more independent, reliable methods within a laboratory*

This procedure is presently more popular than the definitive method procedure, judging from the various certificates of analysis which show that two or more methods have been used to certify the concentrations of most analytes. The procedure has been used extensively by NIST and the NRCC. The methods are usually used within the CRM-producing laboratory, but sometimes a few selected outside laboratories contribute some data. The certified concentrations are usually the means or weighted means of the pooled results, and the uncertainties are expressed as 95% prediction interval plus an allowance for systematic error among the methods used (NIST, 1990a) or 95% tolerance limit for an individual subsample (NBS, 1978). This limit means that 95% of the subsamples from any bottle would give concentrations within the specified range 95% of the time. The NWRI also uses this procedure, in addition to using interlaboratory results, for confirmation, as discussed further in Section 5.3.3.5.

5.3.3.3 *Interlaboratory comparison studies*

This approach is used by several agencies such as the IAEA (Austria), BCR (Belgium), CANMET (Canada), IGGE in collaboration with the IRMA (China), NIES (Japan), RIAP (Russia), and Council for Mineral Technology (South Africa). Each agency uses its own statistical analysis to treat the data leading to certification of the tested material.

The originating agency, such as the IAEA, organizes and distributes the candidate material to the participating laboratories, which report the analytical results from replicate analyses back to the agency within a specified time (e.g., 3 to 6 months). Materials "which have one or more properties sufficiently well established from statistical evaluation of previous interlaboratory comparison studies..." become reference materials with specified reference values and confidence intervals (AQCS, 1994). The BCR also organizes and runs interlaboratory (i.e., round-robin) studies involving expert laboratories of the member states using different methods to acquire data which are then statistically treated to obtain best estimate values (mean of all acceptable data) and uncertainties which are the half-width of the 95% confidence intervals (BCR, 1994; Quevauviller et al., 1994a,b).

CANMET also utilizes the interlaboratory procedure, where a consensus approach is used for certification. Each of the "at least ten" participating laboratories uses its method of choice. Before the consensus values (i.e., usually the mean values) can be calculated, the interlaboratory data must meet a set of certification criteria. A set of results with very high imprecision and the outliers (i.e., results which differ by more than twice the standard deviation of the overall mean) are rejected. A one-way ANOVA is then used to compute the consensus value and its variance (Steger, 1980, 1984).

The IGGE initiated the preparation of eight stream sediment reference materials (GSD 1 to 8) in 1978. The collaborative study started in 1980–81 with 41 participating laboratories in China using different methods. The sediments were certified in 1983, and the recommended values for 50 minor and trace elements were derived by a method of repetitive elimination of outliers and calculation of a central "tendency" from several central values (Xie et al., 1985a). Three other stream sediments were prepared and certified in 1989 by the IGGE in collaboration with the IRMA (Xie et al., 1989). Forty-five laboratories participated in the last certification exercise, which resulted in 72 certified values for major, minor, and trace elements. The IRMA also certified a marine sediment, GSMS-1 (Wang, 1994).

The NIES issued a certified pond sediment reference material in 1981. Each certified value was based on interlaboratory results from at least three independent methods used by 20 to 30 qualified participating laboratories (NIES, 1981; Iwata et al., 1983a,b; Okamoto, 1994; Okamoto and Fuwa, 1985). The uncertainty was estimated based on two standard deviations of the mean of acceptable values and the 95% confidence intervals for the mean of individual methods.

The Council for Mineral Technology (MINTEK) of the South Africa Bureau of Standards had 19 laboratories in five countries participating in the analytical program. Each laboratory uses its own method. A certified value is assigned to a constituent using the Gastwirth median of a data set which has met four predetermined criteria and which has a slightly skewed distribution (Ring, 1993).

RIAP had 32 Soviet, 1 Czechoslovakian, and 1 Bulgarian laboratories participating in the interlaboratory study to certify three marine sediments, SDO-1, SDO-2, and SDO-3 (Berkovits and Lukashin, 1984; Berkovits et al., 1984, 1991). Each certified value was determined as a weighted average, with the 95% confidence level estimated on the basis of Chebyshev's inequality. For certification of other sediments (the OO series), 113 laboratories participated (Berkovits et al., 1991).

The interlaboratory approach is a multilaboratory, multimethodology approach, where the certified values are obtained based solely or mainly on intercomparison results. A variation would be the use of a multilaboratory, single definitive method approach.

5.3.3.4 Select laboratories method

Several authors recognize that subjective judgment is on occasion chosen over rigorous statistical considerations and that the interlaboratory approach has its pitfalls, namely, interlaboratory factors, which cause discrepancies in analytical data (De la Roche and Govindaraju, 1969; Ingamells, 1978; Steele et al., 1978; Morrison, 1980). The select laboratories method, introduced by Abbey (1970) and continually refined by Abbey (1983), addresses these factors in a constructive manner. The method is based on three principles, which are, according to Abbey (1983):

1. All results for a given constituent in a given sample are classified as "good," "fair," or "poor" on the basis of their position.
2. Each contributing laboratory is given a rating, based on its relative numbers of good, fair, and poor results.
3. Only the results reported by laboratories with ratings above a specified level are used in arriving at the desired value.

This procedure reduces the original data to a relatively small group of selected results, presumably from good laboratories, thus minimizing the systematic errors which are often prominent in interlaboratory studies. Xie et al. (1989) used this method to confirm their best estimated values calculated from several central values.

5.3.3.5 "Quasi-interlaboratory" procedure

At the NWRI, the interlaboratory studies form part of the certification process, but their results are used for confirmation only, to provide laboratory performance, or as an additional data set to the pooled data, the bulk of which comes from in-house analyses and selected, contracted laboratories of known performance. A minimum of two independent, reliable methodologies must be used, and the compatibility of the methodology must be no less than 10% most of the time (Cheam and Chau, 1984; Lee et al., 1986, 1987; Lee and Chau, 1987; Cheam et al., 1989; Stokker and Kaminski, 1995). In certain cases, such as Se determination, the compatibility of three independent methodologies was high (i.e., 25%, resulting from three independent groups of results of 1.02, 1.11, and 1.31 μg/g of Se). However, it was considered acceptable due to both the low level of Se and the recognized difficulty of Se analysis (Cheam et al., 1989). In the certification of metals in sediments, a methodology refers to a group of methods that use one type of instrument/spectrometer; for example, atomic absorption spectrometric methodology can include several different methods using an atomic absorption spectrometer as the base instrument of detection, whereas GC-FID and GC-MS are taken as two different methods in the determination of organic compounds (Lee et al., 1987), and GC-ECD/ultrasonic extraction and GC-ECD/Soxhlet extraction under various conditions are considered to be different methods (Lee and Chau, 1987).

5.3.3.6 Single method, single laboratory

Terashima et al. (1990, 1992) of the Geological Survey of Japan used one method, presumably in-house, to certify a series of sedimentary rocks, including four aquatic sediments.

5.4 Discussion, recommendations, and conclusions

5.4.1 Role of certified reference materials

The role of CRMs has been previously discussed at length by several authors (Cali et al., 1975; Uriano and Gravatt, 1977; Taylor, 1985). CRMs are essentially used for:

1. Intralaboratory QC to ensure accuracy of analyses
2. Calibration of an instrument or a measurement process
3. The primary standard for determination of a chemical
4. Interlaboratory QC to determine method accuracy and method compatibility and to help establish certified values for CRM candidates
5. Method development/evaluation to ascertain method accuracy
6. Development of new CRMs or secondary reference materials
7. Direct primary calibration in the field (Uriano and Gravatt, 1977)

5.4.2 Sediment certified reference materials

Table 5-1 lists the sediment reference materials which have recommended trace metal values reported in a recent extensive compilation of 383 geostandards by Govindaraju (1994), as well as other sediment CRMs not listed in the compilation. For each CRM, the sediment type, name, producer, origin, and a key reference are given. Table 5-2 summarizes the recommended values for the 13 U.S. Environmental Protection Agency priority elements for each of the CRMs listed in Table 5-1. The major, minor, and other trace elements are too numerous to be included here.

Table 5-3, following the same format as Table 5-1, shows sediment CRMs for organic compounds which have recommended values reported in a large compilation report by Cantillo (1992), as well as in the latest information sheets and journal publications. Table 5-4 gives the recommended values for each of the certified analytes and for each CRM listed in Table 5-3. Organometallics are also incorporated in these two tables.

Table 5-5 lists the sediment CRMs for radionuclides, again using the format of Table 5-1. Table 5-6 gives the recommended values for each of the certified analytes and for each CRM listed in Table 5-5.

5.4.3 Sediment certified reference materials in preparation

During the preparation of this chapter, several sediment samples were, or will be, prepared for future CRMs. The following is a list of some of them:

Table 5-1 Sediment Certified Reference Materials for Inorganic Parameters

CRM type	Name	Producer[a]	CRM origin	Reference
Anomalous sediment	OOKO201	RIAP	Yakut SSR (N-E part): Composite of two portions of loose sediments, subsoil horizon (Yakutia)	Petrov et al., 1988
	OOKO204	RIAP	Yakut SSR (N-E part): Composite of three portions of loose sediments, subsoil horizon (Yakutia)	Petrov et al., 1988
Background sediment	OOKO202	RIAP	Yakut (N-E part): Subsoil horizon of Aldan Region, Yakutia (fine sand-clay)	Petrov et al., 1988
	OOKO203	RIAP	Yakut (N-E part): Subsoil horizon of Oymyakon region, Yakutia (fine sand-clay)	Petrov et al., 1988
Coastal sediment	IAEA-356	IAEA	"Hot spot," polluted marine sediment	AQCS, 1994, 1995
Estuarine sediment	CRM 277	BCR	Lyophilized sediment from the Scheldt Estuary	BCR, 1994
	BEST-1	NRCC	Sediment from the Beaufort Sea (Hg only)	NRCC, 1990
	MESS-1	NRCC	Sediment from the Miramichi Estuary, Gulf of St. Lawrence	NRCC, 1981
	NBS1646	NIST	Dredged sediment from Chesapeake Bay	NBS, 1982
Lake sediment	CRM 280	BCR	Lyophilized sediment from Lake Maggiore	BCR, 1992
	Jlk-1	GSJ	Lake Biwa, freshwater lake sediment (sedimentary rock series), Shiga Prefecture, Japan	Terashima et al., 1990
	LKSD-1	CCRMP	Joe Lake and Brady Lake, Ontario (Canadian Shield: National Topographic System 31F and 31M)	Lynch, 1990
	LKSD-2	CCRMP	Composite sample: Calabogie Lake, Ontario, and two locations from Northwest Territories, Canada	Lynch, 1990
	LKSD-3	CCRMP	Composite sample: Calabogie Lake, two locations from Manitoba and six locations from Ontario	Lynch, 1990
	LKSD-4	CCRMP	Big Gull Lake, Ontario, Key and Sea Horse Lake, Saskatchewan, Canada	Lynch, 1990
	SL-1	IAEA	Sardis Reservoir Panola County, Mississippi	AQCS, 1994, 1995
	SL-3	IAEA	Sardis Reservoir Panola County, Mississippi	AQCS, 1994, 1995

	WQB-1	NWRI	Lake Ontario sediment, Canada	Cheam et al., 1984
	WQB-3	NWRI	Sediment mixture from Hamilton Harbor (53 kg) and Lake Ontario (160 kg), Canada	Cheam et al., 1989
Marine mud	MAG-1	USGS	Very fine grey-brown bottom sediment from the Wilkinson Basin of the Gulf of Maine	Manheim et al., 1976
	GBW 07313	NRCCRM	Central part of Pacific, collected during the HY4-871 cruise of R/V Haiyang of the MGMR, China	IMG, 1990
	SD-M-2/TM	IAEA	Lyophilized, produced in cooperation with the regional seas program of UNEP	AQCS, 1994, 1995
	MESS-2	NRCC	Estuarine sediment from the Beaufort Sea	NRCC, 1994
	BCSS-1	NRCC	Estuarine sediment from the Bay des Chaleurs, Gulf of St. Lawrence	NRCC, 1994
	PACS-1	NRCC	Harbor sediment from Esquimalt Harbor in British Columbia, Canada	NRCC, 1994
	GSMS-1	IRMA	Central part of Pacific, cruise HY3-871, conducted by Haiyang IV, People's Republic of China	Wang, 1994
	SDO-1	RIAP	Eastern Pacific, 2962 m deep, dark grey and reduced viewed as terrigenous clay	Berkovits and Lukashin, 1984; Berkovits et al., 1991
	SDO-2	RIAP	Central Pacific near Hawaiian Islands, 4680 m deep, oxidized deep-sea clay	Berkovits and Lukashin, 1984; Berkovits et al., 1991
	SDO-3	RIAP	Red Sea, 1350 m deep, oxidized white-colored, calcareous ooze	Berkovits and Lukashin, 1984; Berkovits et al., 1991
Pond sediment	NIES 2	NIES	Surface sediment of Sanshiro Pond within university grounds, University of Tokyo, Japan	NIES, 1981
River sediment	CRM 320	BCR	Lyophilized sediment from the River Toce	BCR, 1994
	GBW 08301	NRCCRM	No information given	NRCCRM, 1992
	GSD-9	IGGE	River sediment in Yangtze River upstream from Wulham, Hubei Province, China	IGGE, 1986

Table 5-1 Sediment Certified Reference Materials for Inorganic Parameters (continued)

CRM type	Name	Producer[a]	CRM origin	Reference
River sediment (continued)	NBS1645	NIST	Dredged material from bottom of Indiana Harbor Canal near Gary, Indiana	NBS, 1978
	NBS2704	NIST	Collected from the bottom of the Buffalo River, by Ohio Street Bridge, Buffalo, New York	NIST, 1990a
Stream sediment	GBW 07309	NRCCRM	No information given	NRCCRM, 1992
	GBW 07310	NRCCRM	No information given	NRCCRM, 1992
	GBW 07311	NRCCRM	No information given	NRCCRM, 1992
	GBW 07312	NRCCRM	No information given	NRCCRM, 1992
	GSD-1	IGGE-IRMA	Tributary flowing through the biotite granite in Shanxi Province, China	IGGE, 1986
	GSD-2	IGGE–IRMA	Tributary flowing through the biotite granite in Jiangxi Province, China	IGGE, 1986
	GSD-3	IGGE–IRMA	Mixture of two sources from Jiangxi, China: porphyry deposit and phyllite rocks	IGGE, 1986
	GSD-4	IGGE–IRMA	From area of carbonate rocks in Anhui Province, China	IGGE, 1986
	GSD-5	IGGE–IRMA	Pond sediment from area of diorite, carbonate rocks in Anhui Province, China	IGGE, 1986
	GSD-6	IGGE–IRMA	From high and cold mountainous terrain of diverse rocks in Qinghai Province, China	IGGE, 1986
	GSD-7	IGGE–IRMA	From lead–zinc area; drainage basin: mica schist, in Liaoning Province, China	IGGE, 1986
	GSD-8	IGGE–IRMA	From a region with acidic to intermediate volcanic rocks in Quangxi, China	IGGE, 1986
	GSD-9	IGGE–IRMA	River sediment in Yangtze River upstream from Wulham, Hubei Province, China	IGGE, 1986
	GSD-10	IGGE–IRMA	From a tributary draining carbonate rocks in Yishan, Guangxi Province, China	IGGE, 1986

GSD-11	IGGE–IRMA	Various rocks from the Shizhuyuan ore field in Binzhou, Hunan Province, China	IGGE, 1986
GSD-12	IGGE–IRMA	From a tributary draining various rocks in Yangchun ore field Guangdong, China	IGGE, 1986
SARM46	MINTEK	Prepared by National Institute of Metallurgy and South African Geological Survey	Ring, 1993
SARM51	MINTEK	Prepared by National Institute of Metallurgy and South African Geological Survey	Ring, 1993
SARM52	MINTEK	Prepared by National Institute of Metallurgy and South African Geological Survey	Ring, 1993
JSd-1	GSJ	Composite sample of northern region, Ibaraki Prefecture, Japan	Terashima et al., 1990
JSd-2	GSJ	Composite sample of eastern region, Ibaraki Prefecture, Japan	Terashima et al., 1990
JSd-3	GSJ	Composite sample of central region, Ibaraki Prefecture, Japan	Terashima et al., 1990
STSD-1	CCRMP	Lavant Creek, Ontario, Canada (31F in National Topographic System)	Lynch, 1990
STSD-2	CCRMP	Hirok Stream Composite Sample 4, British Columbia, Canada (104F, 93A, 93B in NTS)	Lynch, 1990
STSD-3	CCRMP	Hirok Stream and Lavant Creek Composite Sample 4, Canada (104P, 31F, 93A, 93B in NTS)	Lynch, 1990
STSD-4	CCRMP	Composite Sample 5 and 4, Canada (31F, 93A, 93B in NTS)	Lynch, 1990

[a] BCR = Community Bureau of Reference (now Measurements and Testing Program), Commission of the European Community, Belgium; CCRMP = Canada Centre for Mineral and Energy Technology, Certified Reference Material Project, Ottawa, Canada; GSJ = Geological Survey of Japan, Yatabe, Tsukuba, Ibaraki, 305 Japan; IAEA = International Atomic Energy Agency, Analytical Quality Control Services, Vienna, Austria; IGGE = Institute of Geophysical and Geochemical Exploration, Ministry of Geology, Beijing, China; IRMA = Institute of Rock and Mineral Analysis, Ministry of Geology and Mineral Resources, People's Republic of China; MINTEK = Council for Mineral Technology, SABS, Pretoria, South Africa; NIES = National Institute for Environmental Studies, Yatabe-machi, Japan; NIST = National Institute of Standards and Technology, Gaithersburg, Maryland; NRCC = National Research Council of Canada; NRCCRM = National Research Center for CRMs, Beijing, China; NWRI = National Water Research Institute, Burlington, Ontario, Canada; RIAP = Research Institute of Applied Physics, Irkutsk State University, Russia; USGS = U.S. Geological Survey, Denver, Colorado.

Table 5-2 Certified Values for U.S. EPA Priority Elements in Sediment Certified Reference Materials

CRM type	CRM name	Producer[b]	Ag	As	Be	Cd
		Recommended values for EPA priority elements (µg/g)[a]				
Anomalous sediment	OOKO201	RIAP	35	8000	4	9
	OOKO204	RIAP	2.3	60	~3	3
Background sediment	OOKO202	RIAP	0.06	—	2	—
	OOKO203	RIAP	0.1	40	2.1	—
Coastal sediment	IAEA-356	IAEA	8.4	26.9	—	4.47
Estuarine sediment	CRM 277	BCR	—	47.3	—	11.9
	BEST-1	NRCC	—	—	—	—
	MESS-1	NRCC	—	10.6	1.9	0.59
	NBS1646	NIST	—	11.6	1.5	0.36
Lake sediment	CRM 280	BCR	~1.2	51	~3.0	1.6
	JLk-1	GSJ	~0.205	27.7	2.9	1.5
	LKSD-1[c]	CCRMP	0.6	40	1.1	—
	LKSD-2[c]	CCRMP	0.8	11	2.5	—
	LKSD-3[c]	CCRMP	2.7	27	1.9	—
	LKSD-4[c]	CCRMP	—	16	1	—
	SL-1	IAEA	—	27.5	—	0.26
	SL-3	IAEA	—	3.2	—	—
	WQB-1	NWRI	—	23	—	~2.1
	WQB-3	NWRI	—	18.8	—	~3.85
Marine mud	MAG-1	USGS	~0.08	~9.2	3.2	0.202
Marine sediment	GBW 07313	NRCCRM	—	5.8	—	—
	SD-M-2/TM	IAEA	—	18.3	—	113
	MESS-2	NRCC	0.18	20.7	2.32	0.24
	BCSS-1	NRCC	—	11.1	1.3	0.25
	PACS-1	NRCC	—	211	—	2.38
	GSMS-1	IRMA	—	5.8	—	—
	SDO-1	RIAP	—	—	1.4	—
	SDO-2	RIAP	—	—	1.8	—
	SDO-3	RIAP	—	—	0.1	—
Pond sediment	NIES 2	NIES	—	12	—	0.82
River sediment	CRM 320	BCR	—	76.7	~2.5	0.533
	GBW 08301	NRCCRM	—	56	~3.5	2.45
	GSD-9	IGGE	0.089	8.4	1.8	0.26
	NBS1645	NIST	~1.76	66	~0.75	10.2
	NBS2704	NIST	—	23.4	—	3.45
Stream sediment	GBW O7309	NRCCRM	0.089	8.4	1.8	0.26
	GBW O7310	NRCCRM	0.27	25	0.9	1.12
	GBW O7311	NRCCRM	3.2	188	26	2.3
	GBW O7312	NRCCRM	1.15	115	8.2	4
	GSD-1	IGGE-IRMA	0.048	1.96	3	0.088
	GSD-2	IGGE-IRMA	0.066	6.2	17.1	0.065
	GSD-3	IGGE-IRMA	0.59	17.6	1.5	0.1
	GSD-4	IGGE-IRMA	0.084	19.7	2.4	0.19
	GSD-5	IGGE-IRMA	0.36	75	2.3	0.82

Table 5-2 Certified Values for U.S. EPA Priority Elements in Sediment Certified Reference Materials (continued)

Recommended values for EPA priority elements (μg/g)[a]								
Cr	Cu	Hg	Ni	Pb	Sb	Se	Tl	Zn
75	250	—	24	100	150	~2	1	390
29	240	—	18	150	~30	~0.2	~2	140
65	44	—	31	14	~0.4	~0.2	~0.6	54
120	50	—	55	15	~5	~0.2	~0.4	86
69.8	365	7.62	36.9	3.47	8.33	—	—	977
192	101.7	1.77	43.4	146	~4	2.04	—	547
—	—	0.092	—	—	—	—	—	—
71	25.1	—	29.5	34	0.73	0.34		191
76	18	0.063	32	28.2	0.4	0.6	0.5	138
114	70.5	0.67	73.6	80.2	~1.4	0.68	~0.7	291
69	59.8	—	36.9	45	—	—	1.11	1.51
31	44	—	16	82	1.2	—	—	331
57	37	—	26	44	1.1	—	—	209
87	35	—	47	29	1.3	—	—	152
33	31	—	31	91	1.7	—	—	194
104	30	—	44.9	37.7	1.31	—	—	223
—	—	—	—	—	0.56	—	—	—
—	~80	1.09	~59.3	~85.7	—	1.02	—	~279
—	~83.4	2.75	52	~243	—	1.15	—	1396
97	30	~0.017	53	24	0.96	1.16	~0.59	130
58.4	424	—	150	29.3	1.85	—	—	160
77.2	32.7	54	56.1	22.8	0.99	—	—	74.8
106	39.3	0.092	49.3	21.9	1.09	0.72	~0.98	172
123	18.5	—	55.3	22.7	0.59	0.43	~0.6	119
113	452	4.57	44.1	404	171	1.09	—	824
58.4	424	—	150	29.3	1.83	—	—	160
62	160	—	190	—	—	—	—	260
240	180	—	150	—	—	—	—	130
31	33	—	35	—	—	—	—	110
75	210	~1.3	40	105	~2	—	—	343
138	44.1	1.03	75.2	42.3	~0.6	0.214	~0.5	142
90	53	0.22	~32	79	—	0.39	—	~251
85	32.1	0.083	32.3	23	0.81	0.16	0.49	78
3	109	1.1	45.8	714	51	1.5	1.44	1720
135	98.6	1.47	44.1	161	3.79	1.12	1.06	438
85	32.1	0.083	32.3	23	0.81	0.16	0.49	78
136	22.6	0.28	30.2	27	6.3	0.28	0.21	46
40	78.6	0.072	14.3	636	14.9	0.2	2.9	373
35	1230	0.056	12.8	285	24.3	0.25	1.76	498
194	21.8	0.018	76	24.4	0.22	~0.11	0.61	79
12.2	4.9	0.04	5.5	32	0.46	~0.21	1.9	44
87	177	0.05	25.6	40	5.4	1.06	0.58	52
81	37.3	0.044	40	30.4	1.84	~0.28	1.2	101
70	137	0.1	34	112	3.9	~0.26	1.16	243

Table 5-2 Certified Values for U.S. EPA Priority Elements in Sediment Certified Reference Materials (continued)

		Recommended values for EPA priority elements (μg/g)[a]				
CRM type	CRM name	Producer[b]	Ag	As	Be	Cd
Stream sediment (continued)	GSD-6	IGGE-IRMA	0.36	13.6	1.7	0.43
	GSD-7	IGGE-IRMA	1.05	0.84	2.7	1.05
	GSD-8	IGGE-IRMA	0.062	2.4	2	0.081
	GSD-9	IGGE-IRMA	0.089	8.4	1.8	0.26
	GSD-10	IGGE-IRMA	0.27	25	0.9	1.12
	GSD-11	IGGE-IRMA	3.2	188	26	2.3
	GSD-12	IGGE-IRMA	1.15	118	8.2	4
	SARM46	MINTEK	—	—	—	—
	SARM51	MINTEK	—	—	—	—
	SARM52	MINTEK	—	—	—	—
	JSd-1	GSJ	~0.036	2.36	1.3	1148
	JSd-2	GSJ	~1.04	39.4	0.8	3580
	JSd-3	GSJ	~3.01	261	10	6175
	STSD-1c	CCRMP	—	23	1.6	—
	STSD-2[c]	CCRMP	0.5	42	5.2	—
	STSD-3[c]	CCRMP	—	28	2.6	—
	STSD-4[c]	CCRMP	—	15	1.7	—

[a] ~ = proposed or for information only value; — = value not given.

[b] BCR = Community Bureau of Reference (now Measurements and Testing Program), Commission of the European Community, Belgium; CCRMP = Canada Center for Mineral and Energy Technology, Certified Reference Material Project, Ottawa, Canada; GSJ = Geological Survey of Japan, Yatabe, Tsukuba, Ibaraki, 305 Japan; IAEA = International Atomic Energy Agency, Analytical Quality Control Services, Vienna, Austria; IGGE = Institute of Geophysical and Geochemical Exploration, Ministry of Geology, Beijing, China; IRMA = Institute of Rock and Mineral Analysis, Ministry of Geology and Mineral Resources, People's Republic of China; MINTEK = Council for Mineral Technology, SABS, Pretoria, South Africa; NIES = National

- NIES No. 12: Marine sediment for organotin compounds, information sheet 1995 (Okamoto, 1994)
- NIES No. 16: River sediment (1995 National Institute for Environmental Studies information sheet/J. Yoshinga)
- GSMS-2 and GSMS-3: Marine sediments, by IRMA (Govindaraju, 1994)
- IAEA-383: Coastal sediment for sterols, fecal (AQCS, 1994)
- IAEA-315: Marine sediment, Arabian Sea, for transuranics, γ and β emitters (AQCS, 1994)

5.4.4 Availability

The following producers of sediment CRMs can be contacted for information or to purchase or obtain CRMs free of charge:

Table 5-2 Certified Values for U.S. EPA Priority Elements in Sediment Certified Reference Materials (continued)

Recommended values for EPA priority elements (μg/g)[a]								
Cr	Cu	Hg	Ni	Pb	Sb	Se	Tl	Zn
190	383	0.045	78	27	1.25	~0.3	1.08	144
122	38	0.053	53	350	2.6	~0.31	0.93	238
7.6	4.1	0.042	2.7	21	0.24	~0.15	0.78	43
85	32.1	0.083	32.3	23	0.81	0.16	0.49	78
136	22.6	0.28	30.2	27	6.3	0.28	0.21	46
40	78.6	0.072	14.4	636	14.9	0.2	2.9	373
35	1230	0.056	12.8	285	24.3	0.25	1.8	498
559	566	—	~125	~1300	—	—	—	5900
509	268	—	178	5200	—	—	—	2200
1300	219	—	182	1200	—	—	—	264
22	22.2	—	6.9	14	—	—	0.404	99
104	1114	—	94	151	—	—	—	2070
35	4.26	—	19.6	82	—	—	—	139
67	36	—	24	35	3.3	—	—	178
116	47	—	53	66	4.8	—	—	246
80	39	—	30	40	4	—	—	204
93	65	—	30	18	7.3	—	—	107

Institute for Environmental Studies, Yatabe-machi, Japan; NIST = National Institute of Standards and Technology, Gaithersburg, Maryland; NRCC = National Research Council of Canada, Ottawa, Canada; NRCCRM = National Research Center for CRMs, Beijing, China; NWRI = National Water Research Institute, Burlington, Ontario, Canada; RIAP = Research Institute of Applied Physics, Irkutsk State University, Russia; USGS = U.S. Geological Survey, Denver, Colorado.

[c] Recommended values are those from Govindaraju (1994). However, those given by Lynch (1990) and Bowman (1990) are provisional.

Canada Centre for Mineral and Energy Technology (CANMET) Certified Reference Materials Project (CCRMP)
555 Booth Street, Ottawa, Ontario K1A 0G1, Canada

Community Bureau of Reference (BCR, now Measurements and Testing Programme)
Commission of the European Community, Rue de la Loi 200, B-1049 Brussels, Belgium

Council for Mineral Technology (MINTEK)
South Africa Bureau of Standards, Private Bag X191, Pretoria, Transvaal 0001, Republic of South Africa

Geological Survey of Japan (GSJ)
Geochemistry Department, 1-1-3 Higashi, Yatabe, Tsukuba, Ibaraki 305, Japan

Table 5-3 Sediment Certified Reference Materials for Organic Parameters

CRM type	Name	Producer[a]	Origin	Reference
Coastal sediment	CRM 462	BCR	Arcachon Bay, France	BCR, 1994
	IAEA-356	IAEA	"Hot spot," polluted marine sediment for MeHg	AQCS, 1995
	IAEA-357	IAEA	"Hot spot," polluted marine sediment collected from Lagoon of Venice, Italy	AQCS, 1994, 1995
Lake sediment	DX-1	NWRI	Great Lakes sediment	NWRI, 1995a
	DX-2	NWRI	Great Lakes sediment	NWRI, 1995a
	EC-1	NWRI	Hamilton Harbor, Ontario, Canada	Lee et al., 1987
	EC-2	NWRI	Lake Ontario, near Niagara River, Ontario, Canada	Lee et al., 1986
	EC-3	NWRI	Niagara River plume in Lake Ontario, Canada	NWRI, 1992e,f,g
Marine sediment	CS-1	NRCC	Laurential Channel midway between Nova Scotia and Newfoundland, Canada	NRCC, 1982
	HS-1	NRCC	Organic- and sulfur-rich sediment from Nova Scotia Harbor, Canada	NRCC, 1982
	HS-2	NRCC	Organic- and sulfur-rich sediment from Nova Scotia Harbor, Canada	NRCC, 1982
	PACS-1	NRCC	Harbor sediment from Esquimalt Harbor in British Columbia, Canada	NRCC, 1990
	SRM 1941	NIST	Chesapeake Bay at the mouth of Baltimore (Maryland) Harbor	NIST, 1989
River sediment	SRM 1939	NIST	Hudson River, New York	NIST, 1990b

[a] BCR = Community Bureau of Reference, Commission of the European Community, Belgium; IAEA = International Atomic Energy Agency, Analytical Quality Control Services, Vienna, Austria; NIST = National Institute of Standards and Technology, Office of Standard Reference Materials, Gaithersburg, Maryland; NRCC = National Research Council of Canada, Ottawa, Canada; NWRI = National Water Research Institute, Burlington, Ontario, Canada.

Table 5-4 Certified Values for Organics in Sediment Certified Reference Materials

CRM type	Name	Producer[a]	Certified compounds (and values)
Coastal sediment	CRM 462	BCR	Di- and tributyltin (as dibutyltin = 128 ± 16 μg/g, as tributyltin = 70 Å 14 μg/kg)
	IAEA-356	IAEA	Methyl mercury (as Hg = 5.46 μg/g)
	IAEA-357	IAEA	PAHs, PCBs, and other organics — see certified compounds and values below
Lake sediment	DX-1	NWRI	Chlorinated dioxins and furans — see certified compounds and values below
	DX-2	NWRI	Chlorinated dioxins and furans — see certified compounds and values below
	EC-1	NWRI	PAHs and PCBs — see certified compounds and values below
	EC-2	NWRI	Chlorobenzenes, hexachlorobenzenes, PAHs — see certified compounds and values below
	EC-3	NWRI	PAHs and chlorobenzenes — see certified compounds and values below
Marine sediment	CS-1	NRCC	PCBs (as Aroclor 1254 = 1.15 ± 0.60 μg/kg)
	HS-1	NRCC	PCBs (as Aroclor 1254 = 21.8 ± 1.1 μg/kg) — see certified compounds and values below
	HS-2	NRCC	PCBs (as Aroclor 1254 = 111.8 ± 2.5 μg/kg) — see certified compounds and values below
	PACS-1	NRCC	Tributyltin (1.27 ± 0.22), dibutyltin (1.16 ± 0.18), and monobutyltin (0.28 ± 0.17) (all in μg/g of Sn)
	SRM 1941	NIST	PAHs: phenanthrene, anthracene, pyrene, fluoranthene — see certified compounds and values below
River sediment	SRM 1939	NIST	PCB 26 (4.20 ± 0.29 μg/g), PCB 28 (2.21 ± 0.10 μg/g), PCB 44 (1.07 ± 0.12 μg/g)

Table 5-4 Certified Values for Organics in Sediment Certified Reference Materials (continued)

Compound	IAEA-357 (ng/g)	Congener	DX-1 (pg/g)	DX-2 (pg/g)
Anthracene	2100 ± 120	2378-TCDD	263 ± 53	262 ± 51
Aroclor 1254	940 ± 202	Total TCDD	416 ± 121	418 ± 125
Benz[*a*]anthracene	74 ± 1200	12378-PeCDD	22 ± 8	28 ± 14
Benzo[*a*]pyrene	6900 ± 1000	Total PeCDD	226 ± 143	253 ± 150
Benzo[*e*]pyrene	6100 ± 3300	123478-HxCDD	23 ± 7	25 ± 8
Benzo[*ghi*]perylene	5200 ± 1000	123678-HxCDD	77 ± 27	85 ± 33
Chrysene	8900 ± 1500	123789-HxCDD	53 ± 24	58 ± 19
HCB	2.4 ± 0.7	Total HxCDD	669 ± 185	739 ± 218
Heptachlor	1.5 ± 0.6	1234678-HpCDD	634 ± 182	757 ± 320
Indeno[1,2,3-*cd*]pyrene	4900 ± 600	Total HpCDD	1251 ± 361	1486 ± 476
p,p'-DDD	30 ± 6	OCDD	3932 ± 933	4402 ± 1257
p,p'-DDE	25 ± 7	Total PCDD	6490 ± 1309	7294 ± 1733
p,p'-DDT	35 ± 12			
PCB 52	47 ± 4	2378-TCDF[b]	89 ± 44	134 ± 61
PCB 101	73 ± 6	Total TCDF	659 ± 259	975 ± 588
PCB 138	74 ± 16	12378-PeCDF	39 ± 14	46 ± 10
PCB 170	15 ± 5.2	23478-PeCDF	62 ± 32	88 ± 28
Phenanthrene	10400 ± 750	Total PeCDF	790 ± 489	916 ± 351
Phytane	720 ± 280	123478-HxCDF	714 ± 276	825 ± 348
Pyrene	15100 ± 2700	123678-HxCDF	116 ± 37	153 ± 61
Resolved aromatics (μg/g)	130 ± 27	123789-HxCDF[b]	28 ± 42	36 ± 45
Sigma alkanes C14–34	14200 ± 5100	234678-HxCDF[b]	57 ± 36	70 ± 47
		Total HxCDF	1800 ± 809	2111 ± 662
		1234678-HpCDF	2397 ± 796	3064 ± 745
		1234789-HpCDF	137 ± 62	152 ± 84
		Total HpCDF	3567 ± 1165	4068 ± 1306
		OCDF	7122 ± 2406	7830 ± 3087
		Total PCDF	13676 ± 3777	15981 ± 4177

Compound	EC-1 (μg/g)
Anthracene	1.2 ± 0.6
Benz[*a*]anthracene	8.7 ± 1.6
Benzo[*b*]fluoranthene	7.9 ± 1.8
Benzo[*k*]fluoranthene	4.4 ± 1.0
Benzo[*a*]pyrene	5.3 ± 1.3
Benzo[*e*]pyrene	5.3 ± 1.3
Benzo[*ghi*]perylene	4.9 ± 1.3
Fluoranthene	23.2 ± 4.1
Indeno[1,2,3-*cd*]pyrene	5.7 ± 1.2
Phenanthrene	15.8 ± 2.5
Pyrene	16.7 ± 3.9
Total PCBs	2.00 ± 0.15

Compound	EC-3 (ng/g)
Benzo[*a*]pyrene	386 ± 100
Benzo[*e*]pyrene	450 ± 98
Benz[*a*]anthracene	312 ± 56
Fluoranthene	558 ± 92
Phenanthrene	293 ± 66
Pyrene	436 ± 94
1,3,5-Trichlorobenzene	113.6 ± 19.0
1,2,3-Trichlorobenzene	8.9 ± 2.4
1,2-Dichlorobenzene	20.7 ± 6.2
1,3-Dichlorobenzene	105.4 ± 35.0
Hexachlorobenzene	279.0 ± 66.2
Hexachlorobutadiene	61.3 ± 13.8
Pentachlorobenzene	65.4 ± 16.4

Compound	EC-2
1,2,3,4-Tetrachlorobenzene (ng/g)	36.5 ± 4.8
1,2,3,5-Tetrachlorobenzene (ng/g)	5.2 ± 0.7
1,2,4,5-Tetrachlorobenzene (ng/g)	84 ± 9.8
1,2,3-Trichlorobenzene (ng/g)	6.1 ± 1.3
1,2,4-Trichlorobenzene (ng/g)	80.7 ± 10.8
1,3,5-Trichlorobenzene (ng/g)	34.3 ± 5.2
1,2-Dichlorobenzene (ng/g)	18.1 ± 2.5
1,3-Dichlorobenzene (ng/g)	74.7 ± 10.1
1,4-Dichlorobenzene (ng/g)	84.4 ± 17.6
Hexachlorobenzene (ng/g)	200.6 ± 26.4
Hexachlorobutadiene (ng/g)	21.3 ± 3.2
Pentachlorobenzene (ng/g)	48.6 ± 4.8
Benz[*a*]anthracene (μg/g)	1.42 ± 0.51
Benzo[*b*]fluoranthene (μg/g)	2.48 ± 0.86
Benzo[*k*]fluoranthene (μg/g)	1.93 ± 0.72
Benzo[*ghi*]perylene (μg/g)	1.47 ± 0.65
Benzo[*a*]pyrene (μg/g)	1.21 ± 0.56
Benzo[*e*]pyrene (μg/g)	1.91 ± 0.72
Dibenz[*a,h*]anthracene (μg/g)	0.49 ± 0.21
Fluoranthene (μg/g)	3.55 ± 0.83
Indeno[1,2,3-*cd*]pyrene (μg/g)	1.55 ± 0.53
Pyrene (μg/g)	2.92 ± 0.63

Table 5-4 Certified Values for Organics in Sediment Certified Reference Materials (continued)

Congener	HS-1 (μg/kg)	HS-2 (μg/kg)	Compound	SRM 1941 (μg/g)
101	1.62 ± 0.21	5.42 ± 0.34	Phenanthrene	0.577 ± 0.059
138	1.98 ± 0.28	6.92 ± 0.52	Anthracene	0.202 ± 0.042
151	0.48 ± 0.08	1.37 ± 0.07	Pyrene	1.080 ± 0.020
153	2.27 ± 0.28	6.15 ± 0.67	Fluoranthene	1.220 ± 0.240
170	0.27 ± 0.05	1.07 ± 0.15	Benz[*a*]anthracene	0.550 ± 0.079
180	1.17 ± 0.15	3.70 ± 0.33	Benzo[*b*]fluoranthene	0.780 ± 0.190
194	0.23 ± 0.04	0.61 ± 0.07	Benzo[*k*]fluoranthene	0.444 ± 0.049
196	0.45 ± 0.04	1.13 ± 0.12	Benzo[*a*]pyrene	0.670 ± 0.130
199	0.57 ± 0.07	1.39 ± 0.09	Perylene	0.422 ± 0.033
209	0.33 ± 0.10	0.90 ± 0.14	Benzo[*ghi*]perylene	0.516 ± 0.083
			Indeno[1,2,3-*cd*]pyrene	0.569 ± 0.040

[a] BCR = Community Bureau of Reference, Commission of the European Community, Belgium; IAEA = International Atomic Energy Agency, Analytical Quality Control Services, Vienna, Austria; NIST = National Institute of Standards and Technology, Office of Standard Reference Materials, Gaithersburg, Maryland; NRCC = National Research Council of Canada, Ottawa, Canada; NWRI = National Water Research Institute, Burlington, Ontario, Canada.

[b] Provisional values only.

Table 5-5 Sediment Certified Reference Materials for Radionuclides

CRM type	Name	Producer[a]	Origin	Reference
Lake sediment	SL-2	IAEA	Origin not given	AQCS, 1994, 1995
Marine sediment	IAEA-135	IAEA	Collected in June 1991 near Sellafield, Irish Sea, for radionuclides	AQCS, 1994, 1995
	IAEA-300	IAEA	Baltic Sea sediment	AQCS, 1994, 1995
	IAEA-367	IAEA	Collected at nuclear test sites at the Enewetak, Marshall Islands, Pacific Ocean	AQCS, 1994, 1995
	IAEA-368	IAEA	Marine surface sediment from the Lagoon of Mururoa Atoll, Pacific Ocean	AQCS, 1994, 1995
	SD-N-2	IAEA	Origin not given	AQCS, 1994, 1995
Stream sediment	IAEA-313	IAEA	Collected from Sibolga area, west coast of North Sumatra, Indonesia	AQCS, 1994, 1995
	IAEA-314	IAEA	Collected from Kalan area, west Kalimantan (Borneo), Indonesia	AQCS, 1994, 1995
River sediment	SRM 4350	NIST	River sediment	Gladney et al., 1987
	SRM 4350B	NIST	River sediment (1981 certificate of analysis)	NBS, 1981
Lake sediment	SRM 4354	NIST	Gyttja freshwater lake sediment (1986 certificate of analysis)	NBS, 1986

[a] IAEA = International Atomic Energy Agency, Analytical Quality Control Services, Vienna, Austria; NIST = National Institute of Standards and Technology, Gaithersburg, Maryland.

Table 5-6 Certified Radionuclides in Sediment Certified Reference Materials

CRM type	Name	Producer[a]	Certified radionuclides (Bq/kg) for sediment CRMs
Lake sediment	SL-2	IAEA	K-40 = 240, Cs-137 = 2.4
Marine sediment	IAEA-135	IAEA	K-40 = 560, Co-60 = 4.8, Cs-134 = 5.2, Cs-137 = 1108, Eu-154 = 6.8, Eu-135 = 5.5, Ra-226 = 23.9, Ra-228 = 36.7, Th-232(t) = 36.9, Pu-238 = 43, Pu-239+240 = 213
	IAEA-300	IAEA	K-40 = 1059, Co-60 = 1.5, Sb-125 = 11, Cs-134 = 66.6, Cs-137 = 1056.6, Eu-155 = 4.95, Pb-210 = 360, Po-210 = 340.5, Ra-228 = 61.6, U-234 = 69, U-238 = 64.7, Pu-239+240 = 3.55, Am-241 = 1.38
	IAEA-367	IAEA	Co-60 = 1.0, Sr-90 = 102, Cs-137 = 195, Pu-239+240 = 38
	IAEA-368	IAEA	Co-60 = 0.6, Eu-155 = 3.8, Pb-210 = 23.2, Ra-226 = 21.4, Pu-238 = 8.5, U-238 = 31, Pu-239+240 = 31
	SD-N-2	IAEA	K-40 = 220, Cs-137 = 0.8, Th-232 = 4.9, Pu-239+240 = 8.8 mBq/kg
Stream sediment	IAEA-313	IAEA	Ra-226 = 342, Th = 77.1 mg/kg, U = 18.2 mg/kg
	IAEA-314	IAEA	Ra-226 = 732, Th = 17.8 mg/kg, U = 56.8 mg/kg
River sediment	SRM 4350	NIST	Co-60 = 148, Ac-228 = 34, Cs-137 = 100, Eu-154 = 52, K-40 = 540, Mn-54 = 2.1, Pu-239 = 1.4, Sr-90 = 10.3, Zn-65 = 13 (activities as of January 1, 1975)
	SRM 4350B	NIST	Co-60 = 4.64, Cs-137 = 29, Eu-152 = 30.5, Eu-154 = 3.78, Ra-226 = 35.8, Pu-238 = 13 mBq/kg, Pu-239+240 = 0.51, Am-241 = 0.15
Lake sediment	SRM 4354	NIST	Co-60 = 320, Sr-90 = 1090, Cs-137 = 59.2, Th-228 = 28.6, Th-232 = 26.8, U-235 = 0.75, U-238 = 17.4, Pu-238 = 0.26, Pu-239+240 = 4.0, Am-241 = 1.1

[a] IAEA = International Atomic Energy Agency, Analytical Quality Control Services, Vienna, Austria; NIST = National Institute of Standards and Technology, Office of Standard Reference Materials, Gaithersburg, Maryland.

Institute of Geophysical and Geochemical Exploration (IGGE)
Ministry of Geology, Beijing, China (order from Breitlander GMBH, Postfach 8046, D-59035 Hamm, Germany)

Institute of Rocks and Mineral Analysis (IRMA)
Ministry of Geology and Mineral Resources, 26 Baiwanzhuang Road, Beijing 100037, China

International Atomic Energy Agency (IAEA)
Analytical Quality Control Services, Laboratory Seibersdorf, P.O. Box 100, A-1400 Vienna, Austria

National Institute for Environmental Studies (NIES)
Environment Agency of Japan, Yatabe-machi, Tsukuba, Ibaraki, 305, Japan ("reasonable" amounts of NIES CRMs may be obtained on request at no charge from Dr. J. Yoshinaga, Environmental Chemistry Division, National Institute for Environmental Studies, 16-2 Onogawa, Ibaraki 305, Japan)

National Institute of Standards and Technology (NIST)
Office of Standard Reference Materials, Gaithersburg, MD 20899

National Research Center for CRMs (NRCCRM)
Office of CRMs, No. 7 District 11, Hepingjie, Chaoyangqu, Beijing, 100013, China

National Research Council of Canada (NRCC)
Institute for National Measurement Standards, Montreal Road, Ottawa, Ontario, K1A 0R6, Canada

National Water Research Institute (NWRI)
Canada Center for Inland Waters, 867 Lakeshore Road, P.O. Box 5050, Burlington, Ontario, L7R 4A6, Canada

Research Institute of Applied Physics (RIAP)
Irkutsk State University, Russia (can be purchased from ASSO USSR Association of Reference Materials Producers, 4 Krasnoarmeyskaya Street, Sverdiovsk, 620219, USSR, fax (3452) 55 21 78 or from Breitlander, P.O. Box 8046, D-59035 Hamm, Germany, fax (02381) 40 31 89)

U.S. Geological Survey (USGS)
Branch of Geochemistry (Attention: Steve Wilson), Denver Federal Center, Box 25046, MS 473, Denver, CO 80225.

5.4.5 *Recommendations*

1. It is a prerequisite to stock a good selection of pertinent sediment CRMs in any laboratory that analyzes sediments to generate usable analytical data.

2. While it is highly desirable to satisfy the ideal state of matching the matrices and analyte levels (Cheam et al., 1989), it is inconceivable in the near future that such a state can ever be attained. To effectively diversify the matrices and concentrations of analytes, it is recommended that mixing existing CRMs be investigated. This may achieve or approach the ideal state.

5.4.6 *Conclusions*

1. The development and use of CRMs are paramount in an effective QA program.
2. A measurement process that does not use one or more CRMs for QC generates questionable data and thus questionable conclusions and decision making. Unreliable data are more expensive than no data at all.
3. The development of sediment CRMs consumes time and resources and is inefficient. An easy, cost-effective approach for any laboratory would be to mix existing CRMs to produce secondary ones that are tailored to have the desirable matrix and analyte levels.

References

Abbey, S., U.S. Geological Survey standards — a critical study of published analytical data, *Can. Spectrosc.*, 15, 10, 1970.

Abbey, S., An evaluation of USGS III, *Geostand. Newsl.*, 6, 47, 1982.

Abbey, S., Studies in "Standard Samples" of Silicate Rocks and Minerals 1969–1982, Paper 83-15, Geological Survey of Canada, Ottawa, 1983.

Allcott, G.H. and Lakin, H.W., The homogeneity of six geochemical exploration reference samples, in *Geochemical Exploration 1974, Proceedings of the 5th International Geochemical Exploration Symposium,* Special Publication No. 2, Development in Economic Geology, Vol. 1, Elliot, J.L. and Fletcher, W.K., Eds., Elsevier Scientific, New York, 1975, 659.

Analytical Quality Control Services, *Intercomparison Runs References Materials 1992,* International Atomic Energy Agency, Vienna, 1992.

Analytical Quality Control Services, *Analytical Quality Control Services 1994–95,* International Atomic Energy Agency, Vienna, 1994.

Analytical Quality Control Services, *Alternations to the AQCS 1994–95 Catalog,* International Atomic Energy Agency, Vienna, 1995.

Ando, A., Mita, N., and Terashima, S., 1986 values for fifteen GSJ rock reference samples, "igneous rock series," *Geostand. Newsl.*, 11, 159, 1987.

Ando, A., Kamioka, H., Terashima, S., and Itoh, S., 1988 values for GSJ rock reference samples, "igneous rock series," *Geochem. J.*, 23, 143, 1989.

Berkovits, L.A. and Lukashin, V.N., Three marine sediment reference samples: SDO-1, SDO-2 and SDO-3, *Geostand. Newsl.*, 8, 51, 1984.

Berkovits, L.A., Obolyaniniva, V.G., Parshin, A.K., and Romanovskaya, A.R., A system of sediment reference samples: oo, *Geostand. Newsl.*, 15, 85, 1991.

Berman, S.S., Marine sediment reference materials trace metals and inorganic constituents, *Geostand. Newsl.*, 5, 218, 1981.

Berman, S.S., Sturgeon, R.E., and Willie, S.N., Reference materials and the National Research Council of Canada, in *Analytical Quality Control and Reference Materials: Life Sciences, Abstract Book*, Caroli, S. and Morabito, R., Eds., Instituto Superiore di Sanita, Rome, 1994, 6.

Bowman, W.S., Certified Reference Materials CCRMP 90-1E, Canada Centre for Mineral and Energy Technology, Ottawa, Ontario, 1990.

Cali, J.P., The NBS standard reference material program: an update, *Anal. Chem.*, 48, 802A, 1976.

Cali, J.P., Mears, T.W., Michaelis, R.E., Reed, W.P., Seward, R.W., Stanley, C.L., Yolken, H.T., and Ku, H.H., The Role of Standard Reference Materials in Measurement Systems, NBS Monograph 148, National Bureau of Standards, U.S. Department of Commerce, Washington, D.C., 1975.

Cantillo, A.Y., Standards and Reference Materials for Marine Science, NOAA Technical Memorandum NOS ORCA 68, National Oceanic and Atmospheric Administration, Rockville, MD, 1992.

Chau, A.S.Y., Standard reference materials in environmental quality assurance, *Water Qual. Bull.*, 8, 26, 1983.

Chau, A.S.Y. and Lee, H.B., Analytical reference materials. III. Preparation and homogeneity test of large quantities of wet and dry sediment reference materials for long term polychlorinated biphenyl quality control studies, *J. Assoc. Off. Anal. Chem.*, 63, 948, 1980.

Chau, A.S.Y. and Thompson, R., Analytical reference materials. I. Preparation and purification of photomirex, *J. Assoc. Off. Anal. Chem.*, 62, 1302, 1979.

Cheam, V. and Chau, A.S.Y., Analytical reference materials. IV. Development and certification of the first Great Lakes sediment reference material for arsenic, selenium and mercury, *Analyst*, 109, 775, 1984.

Cheam, V., Chau, A.S.Y., and Horn, W., National Interlaboratory Quality Control Study No. 35: Trace Metals in Sediments, Report Series No. 76, Inland Waters Directorate, Burlington, Ontario, 1988.

Cheam, V., Aspila, K.I., and Chau, A.S.Y., Analytical reference materials. VIII. Development and certification of a new Great Lakes sediment reference material for eight trace metals, *Sci. Total Environ.*, 87/88, 517, 1989.

Community Bureau of Reference, *BCR Reference Materials*, Commission of the European Communities, Brussels, 1992.

Community Bureau of Reference, *BCR Reference Materials*, Commission of the European Communities, Brussels, 1994.

De la Roche, H. and Govindaraju, K., Rapport sur deux roches, diorite DR-N et serpentine UB-N, proposées comme étalons analytiques par un groupe de laboratoires français, *Bull. Soc. Fr. Céram.*, 85, 35, 1969.

Dybczynski, R. and Suschny, O., *Final Report on the Intercomparison Run SL-1 for the Determination of Trace Elements in a Lake Sediment Sample*, IAEA/RL/64, International Atomic Energy Agency, Vienna, 1979.

Epstein, M.S., Diamondstone, B.I., and Gills, T.E., A new river sediment standard reference material, *Talanta*, 36, 141, 1989.

Fairbairn, H.W., Schlecht, W.G., Stevens, R.E., Dennen, W.H., Ahrens, L.H., and Chayes, F., A cooperative investigation of precision and accuracy in chemical, spectrochemical, and modal analysis of silicate rocks, *U.S. Geol. Surv. Bull.*, 980, 71, 1951.

Frick, C., The Occurrence and Petrography of the SAROCK Standards, Report, Geological Survey, Pretoria, South Africa, 1981.

Gladney, E.S., O'Malley, B.T., Roelandts, I., and Gills, T.E., Standard Reference Materials: Compilation of Elemental Concentration Data for NBS Clinical, Biological, Geological, and Environmental Standard Reference Materials, NBS Special Publ. 260-111, National Bureau of Standards, U.S. Department of Commerce, U.S. Government Printing Office, Washington, D.C., 1987.

Govindaraju, K., Compilation of working values and sample description for 383 geostandards, *Geostand. Newsl.*, 18, 1, 1994.

Guevremont, R. and Jamieson, W.D., Reference materials for marine trace analysis, *Trends Anal. Chem.*, 1, 113, 1982.

Hathaway, J.C., Data File, Continental Margin Program, Atlantic Coast of the United States, Vol. 2, Sample Collection and Analytical Data, Woods Hole Oceanographic Institute Ref. 71-15, 1971.

Hunter, J.S., The national system of scientific measurement, *Science*, 210, 869, 1980.

Ingamells, C.O., Standard reference materials in geoexploration and extractive metallurgy research, in Geoanalysis '78 Symposium, Ottawa, May 1978.

Institute of Geophysical and Geochemical Exploration, Certificate of Geochemical Standard Reference Materials, Institute of Geophysical and Geochemical Exploration, Langfang, Hebei, China, 1986.

Institute of Marine Geology, Certificate of Certified Reference Material, Marine Sediment, Institute of Marine Geology, Ministry of Geology and Mineral Resources, Qingdao, China, 1990.

International Atomic Energy Agency, Intercomparison run, SL-1, Information sheet, 1977.

International Atomic Energy Agency, Information sheet, December 16, 1980.

International Organization for Standardization, Guide 30, Terms and Definitions Used in Connection with Reference Materials, International Organization for Standardization, Geneva, 1981.

Iwata, Y., Haraguchi, H., Van Loon, J.C., and Fuka, K., Mineralogical characterization of the reference material of "pond sediment," *Bull. Chem. Soc. Jpn.*, 56(2), 434, 1983a.

Iwata, Y., Matsumoto, K., Haraguchi, H., Notsu, K., Okamoto, K., and Fuka, K., Preparation and evaluation of certified reference "pond sediment" (NIES No. 2), *Q. J. Plasma Spectrosc.*, 3(2), 72, 1983b.

Lawrence, J., Chau, A.S.Y., and Aspila, K.I., Analytical quality assurance: key to reliable environmental data, *Can. Res.*, 15, 35, 1982.

Lee, H.B. and Chau, A.S.Y., Analytical reference materials. VII. Development and certification of a sediment reference material for total polychlorinated biphenyls, *Analyst*, 112, 37, 1987.

Lee, H.B., Hong-You, R.L., and Chau, A.S.Y., Analytical reference materials. V. Development of a sediment reference material for chlorobenzenes and hexachlorobutadiene, *Analyst*, 111, 81, 1986.

Lee, H.B., Dookhran, G., and Chau, A.S.Y., Analytical reference materials. VI. Development of a sediment reference material for selected polynuclear aromatic hydrocarbons, *Analyst*, 112, 31, 1987.

Lynch, J., Provisional elemental values for eight new geochemical lake sediment reference materials LKSD-1, LKSD-2, LKSD-3, LKSD-4, and STSD-1, STSD-2, STSD-3, STSD-4, *Geostand. Newsl.*, 14, 153, 1990.

Manheim, F.T., Hathaway, J.C., Flanagan, F.G., and Fletcher, J.D., Marine mud, MAG-1, from the Gulf of Maine, in *Description and Analysis of Eight New USGS Rock Standards*, Flanagan, F.G., Ed., USGS Prof. Paper 840, U.S. Government Printing Office, Washington, D.C., 1976, 25.

Mingcai, Y. et al., The preparation of standard samples of stream sediment, *Geophys. Geochem. Explor.*, 5, 321, 1980.

Moore, L.J. and Machlan, L.A., High accuracy determination of calcium in blood serum by isotope dilution mass spectrometry, *Anal. Chem.*, 44, 2291, 1972.

Morrison, G.H., Analytical controversies, *Anal. Chem.*, 52, 1793, 1980.

Mudroch, A. and Azcue, J., *Manual of Aquatic Sediment Sampling*, Lewis Publishers, Boca Raton, FL, 1995, 219.

National Bureau of Standards, National Bureau of Standards Certificate of Analysis, Standard Reference Material 1645 (River Sediment), Office of Standard Reference Material (Cali, Paul J.), Washington, D.C., 1978.

National Bureau of Standards, National Bureau of Standards Certificate of Analysis, Standard Reference Material 4350B (River Sediment), Office of Standard Reference Material (Uriano, G.A.), Washington, D.C., 1981.

National Bureau of Standards, National Bureau of Standards Certificate of Analysis, Standard Reference Material 1646 (Estuarine Sediment), Office of Standard Reference Material (Uriano, G.A.), Washington, D.C., 1982.

National Bureau of Standards, National Bureau of Standards Certificate of Analysis, Standard Reference Material 4354 (Freshwater Lake Sediment), Office of Standard Reference Material (Uriano, G.A.), Washington, D.C., 1986.

National Institute for Environmental Studies, National Institute for Environmental Studies Certified Material, Pond Sediment Information Sheet, National Institute for Environmental Studies, Environment Agency of Japan, Tsukuba, Ibaraki, Japan, 1981.

National Institute of Standards and Technology, National Institute of Standards and Technology Certificate of Analysis, Standard Reference Material 1941 — Organics in Marine Sediment, National Institute of Standards and Technology, Gaithersburg, MD, 1989.

National Institute of Standards and Technology, National Institute of Standards and Technology Certificate of Analysis, Standard Reference Material 2704 — Buffalo River Sediment, Standard Reference Materials Program (Reed, W.P.), Gaithersburg, MD, 1990a.

National Institute of Standards and Technology, National Institute of Standards and Technology Certificate of Analysis, Polychlorinated Biphenyls Congeners in River Sediment, SRM 1939, National Institute of Standards and Technology, Gaithersburg, MD, 1990b.

National Research Center for CRM, Catalog for Certified Reference Materials (addendum to 1991 catalog), Office of CRMs, Beijing, 1992.

National Research Council of Canada, Marine Analytical Chemistry Standards Program, Marine Sediment Reference Materials MESS-1, BCSS-1 (certificate of analysis), Ottawa, Ontario, 1981.

National Research Council of Canada, Marine Sediment Reference Materials, Polychlorinated Biphenyls (information sheet), Institute for Marine Biosciences, Halifax, Nova Scotia, 1982.

National Research Council of Canada, Marine Analytical Chemistry Standards Program, Reference Materials BCSS-1, MESS-1, PACS-1, BEST-1 (information sheet), Ottawa, Ontario, 1990.

National Research Council of Canada, Marine Analytical Chemistry Standards Program, Environmental Certified Reference Materials Available from NRC (information sheet), Ottawa, Ontario, 1994.

National Water Research Institute, Certified Reference Materials (information sheet), Aquatic Quality Assurance Program, Environment Canada, Burlington, Ontario, 1988.

National Water Research Institute, Certified Reference Material WQB-1, Trace Metals in Sediment, Certificate of Analysis, Environment Canada, Burlington, Ontario, 1990a.

National Water Research Institute, Certified Reference Material WQB-3, Trace Metals in Sediment, Certificate of Analysis, Environment Canada, Burlington, Ontario, 1990b.

National Water Research Institute, Certified Reference Material EC-1, Polychlorinated Biphenyls in Sediment, Certificate of Analysis, Environment Canada, Burlington, Ontario, 1990c.

National Water Research Institute, Certified Reference Material EC-1, Polynuclear Aromatic Hydrocarbons in Sediment, Certificate of Analysis, Environment Canada, Burlington, Ontario, 1992a.

National Water Research Institute, Certified Reference Material EC-2, Polynuclear Aromatic Hydrocarbons in Sediment, Certificate of Analysis, Environment Canada, Burlington, Ontario, 1992b.

National Water Research Institute, Certified Reference Material EC-2, Polychlorinated Biphenyls in Sediment, Certificate of Analysis, Environment Canada, Burlington, Ontario, 1992c.

National Water Research Institute, Certified Reference Material EC-2, Chlorobenzenes in Sediment, Certificate of Analysis, Environment Canada, Burlington, Ontario, 1992d.

National Water Research Institute, Certified Reference Material EC-3, Chlorobenzenes in Sediment, Certificate of Analysis, Environment Canada, Burlington, Ontario, 1992e.

National Water Research Institute, Certified Reference Material EC-3, Polychlorinated Biphenyls in Sediment, Certificate of Analysis, Environment Canada, Burlington, Ontario, 1992f.

National Water Research Institute, Certified Reference Material EC-3, Polynuclear Aromatic Hydrocarbons in Sediment, Certificate of Analysis, Environment Canada, Burlington, Ontario, 1992g.

National Water Research Institute, Certified Reference Materials (information sheet), Certified Reference Values for Chlorinated Dioxins and Furans in Great Lakes Sediments DX-1 and DX-2, Environment Canada, Burlington, Ontario, 1995a.

National Water Research Institute, Quality Assurance Reference Materials & Services, 1995/96 Catalog, National Water Research Institute, Environment Canada, Burlington, Ontario, November 1995b.

Okamoto, K., Biological and environmental reference materials from the National Institute for Environmental Studies, Japan, in *Analytical Quality Control and Reference Materials: Life Sciences, Abstract Book*, Caroli, S. and Morabito, R., Eds., Instituto Superiore di Sanita, Rome, 1994, 5.

Okamoto, K. and Fuwa, K., Certified reference materials at the National Institute for Environmental Studies, *Anal. Sci.*, 1, 206, 1985.

Petrov, L.L., Kornakov, Yu.N., and Persikova, L.A., Catalogue of Reference Samples of Mineral Substance Composition, USSR Academy of Sciences, Siberian Branch, Institute of Geochemistry, Irkutsk, 1988.

Quevauviller, Ph., Certified reference materials for the quality control of environmental analysis within the Measurements and Testing Programme, in *Analytical Quality Control and Reference Materials: Life Sciences, Abstract Book,* Caroli, S. and Morabito, R., Eds., Instituto Superiore di Sanita, Rome, 1994, 4.

Quevauviller, Ph., Astruc, M., Ebdon, L., Desauziers, V., Sarradin, P.M., Astruc, A., Kramer, G.N., and Griepink, B., Certified Reference Material (CRM 462) for the quality control of dibutyl- and tributyl-tin determinations in coastal sediment, *Appl. Organomet. Chem.*, 8, 629, 1994a.

Quevauviller, Ph., Astruc, M., Ebdon, L., Kramer, G.N., and Griepink, B., Interlaboratory study for the improvement of tributyltin determination in harbour sediment (RM 424), *Appl. Organomet. Chem.*, 8, 639, 1994b.

Ring, E.J., The preparation and certification of fourteen South African silicate rocks for use as reference materials, *Geostand. Newsl.*, 19, 137, 1993.

Roelandts, I., Environmental reference materials, *Spectrochim. Acta*, 44B, 925, 1989.

Snedecor, G.W. and Cochran, W.G., Two-way classifications, in *Statistical Methods*, 6th ed., Iowa State University Press, Ames, 1967, chap. 11.

South African Committee for Certified Reference Materials, Catalog for Certified Reference Materials, South Africa Bureau of Standards, Pretoria, 1992.

Steele, T.W., Wilson, A., Goudvis, R., Ellis, P.J., and Radford, A.J., Analyses of the NIMROC Reference Samples for Minor and Trace Elements, Report No. 1945, National Institute of Metallurgy, South Africa, 1978.

Steger, H.F., Certified Reference Materials, CANMET Report 80-6E, Canada Centre for Mineral and Energy Technology, Ottawa, Ontario, 1980.

Steger, H.F., The Canadian Certified Reference Materials Project, in Workshop on Reference Materials, Steger, H.F., Sutarno, R., Bowman, W.S., and Abbey, S., Eds., Energy, Mines and Resources Canada, Ottawa, 1984, 3.

Stokker, Y.D. and Kaminski, E., Development of the first Great Lakes sediment reference materials for chlorinated dioxins and furans, in *DIOXIN '95, 15th Int. Symp. on Chlorinated Dioxins and Related Compounds,* Vol. 23, Bolt, D., Clement, R., Fiedler, H., Harrison, B., Ramamoorthy, S., and Reiner, E., Eds., Dioxin '95 Secretariat, Edmonton, Alberta, Canada, August 21 to 25, 1995.

Sutarno, R. and Faye, G.H., A measure for assessing certified reference ores and related materials, *Talanta*, 22, 675, 1975.

Taylor, J.K., Handbook for SRM Users, NBS Special Publ. 260-100, National Bureau of Standards, U.S. Department of Commerce, U.S. Government Printing Office, Washington, D.C., 1985.

Terashima, S., Ando, A., Okai, T., Kanai, Y., Taniguchi, F., Takizawa, F., and Itoh, S., Elemental concentrations in nine new GSJ rock reference samples, "sedimentary rock series," *Geostand. Newsl.*, 14, 1, 1990.

Terashima, S., Itoh, S., and Ando, A., 1991 Compilation of analytical data for silver, gold, palladium, and platinum in twenty-six GSJ geochemical reference samples, *Bull. Geochem. Surv. Jpn.*, 43, 141, 1992.

Tramontano, J.M., Scudlark, J.R., and Church, T.M., A method for the collection, handling, and analysis of trace metals in precipitation, *Environ. Sci. Technol.*, 21, 749, 1987.

Uriano, G.A., The certified reference materials program of the United States National Bureau of Standards, in *Proc. Int. Symp. on the Production and Use of Reference Materials,* Schmitt, B.F., Ed., Bundesandstalt fur Materialprufung, Berlin, 1980, 7.

Uriano, G.A. and Gravatt, C.C., The role of reference materials and reference methods in chemical analysis, *CRC Crit. Rev. Anal. Chem.*, p. 361, October 1977.

Wang, Y., Certificate of Analysis, Institute of Rock and Mineral Analysis, Beijing, 1994.

Willie, S.N. and Berman, S.S., Certified reference materials, in *Analysis of Contaminants in Edible Aquatic Resources: General Considerations, Metals, Organometallics, Tainting, and Organics*, Kiceniuk, J.W. and Ray, S., Eds., VCH Publishers, New York, 1994, 3.

Xie, X., Yan, M., and Shen, H., Geochemical reference samples, drainage sediments GSD 1-8 from China, *Geostand. Newsl.*, 9, 83, 1985a.

Xie, X., Yan, M., and Shen, H., Usable values for Chinese standard reference samples of stream sediments, soils and rocks: GSD 9-12, GSR 1-6, *Geostand. Newsl.*, 9, 277, 1985b.

Xie, X., Yan, M., Wang, C., and Shen, H., Geochemical reference samples, GSD 9-12, GSS 1-8, GSR 1-6, *Geostand. Newsl.*, 13, 83, 1989.

chapter six

Safety in the laboratory for sediment analysis

José M. Azcue and Alena Mudroch

6.1 Basic rules for laboratory safety

Laboratory safety has been the subject of many articles and books. The objective of this chapter is to summarize some fundamental safety principles for those in laboratories that deal with aquatic sediments. It is beyond the scope of this chapter to fully cover all the relevant topics of safety in the chemical laboratory. However, the references at the end of this chapter should be consulted for detailed information.

Safety of the personnel involved is the most important consideration in any laboratory activity. Working in a laboratory that deals with sediment analysis has an inherent danger similar to that in any other chemical laboratory. Nevertheless, the risks may be minimized by observing proper safety precautions and practices. Most sediment laboratories are equipped with a variety of instruments which are utilized by many users. Multiple users and rotation of personnel can present difficulty in maintaining safe conditions in the laboratory. It should never be assumed that students, technicians, or staff have adequate information on laboratory safety. The manager responsible for the laboratory should provide all users with adequate and updated information to understand hazards in the particular laboratory and their consequences.

The basic safety rules for working in a laboratory where sediment analysis is done can be summarized as follows:

- The laboratory user should be familiar with basic safety rules, such as handling chemicals and laboratory first aid.
- Handle sediment samples and chemicals only in well-ventilated areas and fume hoods.

1-56670-343-3/99/$0.00+$.50

- Gloves and other protective clothing must always be worn when handling sediments; in addition, safety goggles should be worn when necessary.
- Highly chemically and biologically contaminated sediments may require special handling areas and techniques as described in Section 6.4.

All users of the sediment laboratory should be familiar with the labeling system applied to all chemical products. The labeling provides an excellent source of safety information for laboratory chemicals and relevant information on the hazards of chemicals and the safety precautions required, such as universal hazard signs used by the chemical industry. A series of pictograms suggest special dangers or the personal protective clothing recommended for use when handling these substances in the laboratory. Some chemical manufacturers also use color coding for recommended storage of a chemical. The color in the upper left-hand corner of the label is the code for the recommended safe storage condition. For instance, a red label means flammable hazard, and chemicals with such labels should be stored in a flammable liquid storage area (Pipitone, 1984).

Glass is a common material used in the laboratory for sediment analysis. The most complicated equipment can be made of glass. Another advantage of glass is its transparency. However, glass containers have the following three major disadvantages for handling sediments: no resistance to fluorides and hydrofluoric acid, fragility, and susceptibility to thermal shock (Freeman and Whitehead, 1982). The use of breakable containers should be minimized for the storage and transport of sediments. When selecting a piece of glassware for use in sediment handling and analysis, care should be taken to ensure that it is designed for the type of work planned. For pressures even slightly above normal, pressure bottles should be specifically chosen. For filtration with the aid of suction, vacuum flasks should be used. Freeze-drying of sediments under high vacuum is often carried out on samples collected in glass containers. Prior to freeze-drying, sediment samples in the containers have to be frozen. Consequently, the containers should be checked before use to make sure they can withstand the low temperatures which will be used in freezing the samples and the subsequent exposure to high vacuum used in the freeze-dryer. Many types of glass containers can break into small pieces during these operations, and extreme care needs to be taken to remove the broken containers and sediments from the freezer or freeze-dryer. In such cases, safety glasses, gloves made of proper material which will resist cutting by the broken glass, and proper clothing are necessary in the removal of the broken glass and sediment particles. Further, it should be noted that the original volume of wet sediment increases considerably during freezing. Therefore, the containers which will be subject to freezing and freeze-drying should be filled only to two-thirds of their volume to prevent breaking by expansion of the sediment during freezing. Broken or cracked glass should not be placed in waste bins designed to receive other laboratory waste. A separate metal container, appropriately labeled, should be available in each laboratory for the disposal of broken glass.

Sediment samples in the laboratory should always be kept properly stored and labeled. Wet sediments should be stored for a limited time at ±4°C to retard biodegradation and physico-chemical changes (Mudroch and Azcue, 1995). Labels for sediments from contaminated areas should include any available information on the hazard in handling the sediments, such as high levels of cyanide, arsenic, etc. This information is necessary in emergencies and for safe disposal of the sediments after analysis. Efficient management of a laboratory should include periodic inspections of storage containers of chemicals and sediment samples. The objective of the inspection is to detect any deterioration, damage, leakage, or spills from the containers. The inspection should also determine if labels are still legible and adhere firmly to the containers.

The importance of and need to keep a laboratory notebook or protocol are universally accepted among all scientists. Keeping and updating laboratory protocols is one of the simplest and most important tasks performed in the laboratory. However, very few laboratory staff members can affirm that they always keep detailed and complete notes of their activities in the laboratory. Kanare (1985) suggested that writing in the laboratory protocol logbook is an essential part of good science, because it forces the scientist to stop and think about what is being done in the laboratory. All incidents related to chemical hazards or toxicity that happened in the sediment laboratory should be recorded. Therefore, the laboratory notes will help to identify, for example, special handling requirements for sediments or allergic reactions caused by contaminated sediments. Each instrument in the analytical laboratory should also have its own logbook. A standard page of an instrument logbook is shown in Figure 6-1. The instrument logbook generally consists of printed forms to keep chronological records of users, time, type of samples, equipment performance, and technical problems encountered. The logbook should be attached to or kept close to the instrument, and any use of the instrument should always be recorded. Keeping constant and detailed records on the use of the instrument facilitates its maintenance and helps to trace the origin of any problems. All laboratory protocols and notebooks should be bound rather than spiral notebooks or loose-leaf binders. The data in bound notebooks will be recorded in a consistent chronological order, and related notes written on different pages will not be lost accidentally. Industrial laboratories for sediment analyses or researchers whose work may have legal implications or is subject to patents are required to use a bound notebook. Laboratory notebooks should be retained for a length of time set by the government or necessary to fulfill other legal requirements. From a legal standpoint, in the United States, original notes

Laboratory: ____________________

Instrument: ____________________

Date: ____________________

User: ____________________

Time on: ____________________

Type of samples: ____________________

Number of samples: ____________________

Responsible: ____________________

Elements analyzed: ____________________

Gas level: ____________________

Notes: ____

____

____

Time off: ____________________

Figure 6-1 Example of an instrument logbook.

of investigations or discoveries should be available for use as evidence in patent disputes for 23 years after the date a patent is issued (Kanare, 1985).

6.2 *Safety considerations in laboratory design*

6.2.1 *General considerations*

Requirements for the design and construction of new safe laboratories are well described in the literature (e.g., Diberardinis et al., 1987; Fawcett, 1988; Ashbrook and Renfrew, 1991; Furr, 1995). However, very often in the renovation of old laboratories, safety considerations are overlooked or ignored. Therefore, in this section, only the general requirements and measurements that would improve the efficiency and safety of a laboratory for sediment analysis are summarized.

Guidelines for laboratory safety are generally provided by regulatory agencies, such as the National Institute for Occupational Safety and Health (NIOSH) or the Occupational Safety and Health Administration (OSHA). The guidelines provide the framework for a safety strategy that should be further adapted to the specific characteristics of the sediments analyzed (i.e., radioactive materials, biologically contaminated materials, etc.). Each laboratory that deals with contaminated sediments should have its own safety program. The objectives of the safety program are to prevent exposure to hazardous substances and/or injuries. The essential considerations of a safety program include:

- Designating a safety coordinator
- Establishing a training program for personnel working with contaminated sediments
- Maintaining laboratory safety requirements, such as a fire extinguisher, spill and emergency kits, etc.
- Keeping an updated inventory of personnel protection equipment
- Implementing a program for periodic monitoring of laboratory safety conditions (i.e., labels on chemicals, storage of chemicals, etc.)
- Establishing emergency procedures, such as measures to decontaminate workers and notification of the nearest medical facility

6.2.2 *Safety equipment and procedures*

6.2.2.1 *Ventilation and fume hoods*

The laboratory for sediment analysis needs adequate ventilation to provide an environment that is safe and comfortable. The inlet air can be purified by passage through a water curtain and electrostatic precipitators (Freeman and Whitehead, 1982). A good ventilation system prevents the buildup of dangerous concentrations of flammable and toxic gases and transports them out of the laboratory by a system of ductwork and fans. Chemical fume hoods are the most common system of exhaust ventilation found in laboratories for sediment analysis. Diberardinis et al. (1987) defined a chemical fume hood as "the laboratory worker's all-purpose safety device." Sediments should be handled in a fume hood, with the sash pulled down two-thirds to cover the technician's upper body. The fume hood must be made of noncombustible and nonabsorbent materials. The velocity of air entering the fume hood at its face determines whether or not the hood is safe. For a general laboratory, a face velocity of 30 m/min is recommended, while fume hoods for highly toxic materials require face velocities of up to 61 m/min (Steere, 1971). In laboratories with more than one fume hood, separate exhaust systems are required for each unit to avoid back diffusion from one fume hood to another. Special precautions should be taken when working with chemicals that are not detectable by odor or visible, in case they escape from the fume hood. Several devices can be installed to monitor fume hood functions, including liquid-filled draft gauges, maynehelic gauges, and bridled-vane gauges, when working with such chemicals (Diberardinis et al., 1987). The efficiency of the extraction system of the fume hood in dealing with smoke should be tested annually by using a smoke generator with the sash in various positions. To avoid the possibility of auto-ignition of highly flammable solvents, any internal lighting must be flame-proofed and resistant to corrosion; therefore, fluorescent fittings are preferred.

6.2.2.2 *Fire extinguishing systems*

All laboratories should be equipped with fire and smoke detector systems. For sediment laboratories, the use of heat-sensitive detectors is recommended, because ionization detection systems sense combustion products commonly used in the laboratory. Water sprinkler systems are considered the most effective method for fire suppression. Other materials should be used as fire controls in laboratories containing large quantities of materials (e.g., elemental sodium) that react violently with water. Carbon dioxide (CO_2) is the most commonly used fire extinguisher agent in laboratories. All sediment laboratories should be equipped with portable fire extinguishers.

6.2.2.3 *Emergency showers and eyewashes*

The main purpose of emergency showers is to dilute and wash off chemical spills on the human body, not to extinguish clothing fires. Although they can also be used as an effective fire extinguisher for a clothing fire, the best method in such situations is to "stop, drop, and roll" and then remove the burned clothes and seek help (Diberardinis et al., 1987). The proper location and functioning of the emergency showers are very important because many chemicals have strongly toxic effects when in contact with the skin. The shower should be located outside the laboratory, but no farther than 15 m away. One shower in a corridor can serve several laboratories. Low-velocity deluge shower heads are recommended, because high-velocity showers can further damage the skin. The standard requires that the emergency showers be capable of delivering a minimum of about 114 L of water per minute (Pipitone, 1984). The emergency shower should provide water between 21 and 32°C; cold water should be avoided. A viable alternative is to adapt an existing toilet shower to an emergency shower. It is recommended that the injured person remain in the shower for about 15 min and then be immediately transferred to a medical facility for further treatment.

Every sediment laboratory that uses strong acids or bases should have at least one eyewash facility. The eyewash facility should provide copious flushing water of no less than 1.5 L/min, at approximately 20°C. In small and low-risk laboratories, eyewash bottles may be suitable. However, special attention should be given to the location of the bottle to make sure that the wrong bottle cannot be picked up by someone in an emergency. A 15-min wash after a chemical splash to the face and eyes is recommended. Water should not be poured directly on the eyeball but on the surrounding area. Following the 15-min eyewash, the injured person should be transported to a medical facility for further treatment. The standard specifies that emergency eyewashes and showers be activated weekly to flush the line and verify proper operation (Pipitone, 1984). Both eyewash facilities and emergency showers should be accessible within 10 sec.

6.2.2.4 Chemical spill control

Spill of hazardous materials can represent risk of exposure to toxic material, potential for fire, or danger of slipping on a wet surface. Pipitone (1984) provided a checklist for preventing causes of spills; the list includes metal containers free of rust; containers clean and free of any chemical leakage; shelf units well fastened to the wall and not overcrowded; shelves provided with raised edges; and chemicals transported using carts, bottle carriers, or pails. In general, spills of solid materials are easier to clean than those involving liquids. Spilled material can normally be swept by a broom and placed into a suitable waste container. Highly toxic materials may be removed with a vacuum cleaner. Hazardous liquids spilled should be converted into solids so as to be removed mechanically. This may be done by chemical inactivation (neutralization) or using an absorbent or gel (Brugger, 1980; Hedberg, 1982). There are numerous kinds of absorbents, such as minerals (e.g., clay), vegetables (e.g., paper towels and sawdust), and synthetics (e.g., polypropylene fibers) (Brugger, 1980). Every sediment laboratory should be provided with an adequate quantity of neutralizing chemicals, adsorbents, and other materials needed to counteract any accidental spill.

Hydrofluoric acid deserves special attention because of its extreme corrosivity to human skin. A spill of hydrofluoric acid must be treated with a calcium-containing compound and soda ash to precipitate the fluoride ion as harmless calcium fluoride and render a neutral pH. Mercury spills also deserve special attention because of their high toxicity. The most common methods to clean mercury spills are sprinkling powdered sulfur to precipitate mercuric sulfide, specially impregnated granular activated carbon, foam collectors, and recently commercially prepared powders designated to amalgamate the mercury. However, a special mercury monitor should be used after cleanup to ensure absolute decontamination of mercury vapors. All laboratory users should be familiar with the location and use of the cleanup material. The cleanup material needs to be located in an easily accessible place and clearly identified.

6.2.2.5 Emergency cabinets

Emergency cabinets should be provided to all laboratories. General items include an emergency blanket, first aid kit, items specific to the laboratory, and escape breathing equipment for laboratories that work with highly toxic volatile materials. For example, the first aid kit for sediment laboratories that use mercury or cyanide should contain a suitable antidote. The complexity of the emergency cabinet depends on the number of users of the laboratory and the degree of risk of the activities carried out. Laboratories in isolated areas should be better equipped than laboratories with easy access to a hospital.

6.2.2.6 Analytical laboratory

The instrument space should preferably be divided into several separate but contiguous rooms to meet several special needs. It is best to keep most analytical instruments away from the ordinary fumes of a chemical laboratory, particularly acid fumes. Chemical laboratories must have good ventilation and adequate fume hoods and should be supplied with basic utilities such as hot and cold water, distilled water, gas, electric power, and special drains which would not be clogged by sediments. Many small and medium-sized instruments can be placed in one large, general instrument room. These instruments include colorimeters, spectrophotometers, pH meters, titrameters, fluorescence meters, etc. Such apparatus generally requires only basic utilities but special electricity. The laboratory space should preferably be completely air-conditioned and, most importantly, the relative humidity should be kept at about 40 to 50%. This will increase the life of the instruments and will permit working with materials such as NaCl or KBr cells which are needed in the infrared region.

Working space around each instrument should be provided with a cup sink and a cold water outlet, a gas outlet, an air outlet, and several electrical outlets. It is recommended that desks for general instruments be about 60 cm high, so that the instruments can be easily controlled by a seated operator and the dials on the top of the instruments can be clearly read and operated. Space below the desktop should be open to allow for knee space. The desktop can be made of synthetic materials such as Formica but should not be of metal because many electrical instruments will be placed on it. In general, an instrument laboratory should be provided with sufficient storage space for spare parts and repair tools.

Some instruments, either because of their size or special requirements or unusual hazards in their operation, should be isolated in smaller rooms or areas. Due to the immense variety of instruments and equipment employed in a sediment laboratory, it is not feasible to review the hazards and safeguards which should be considered for each laboratory instrument. Technical information should be obtained from the manufacturer before purchasing the instruments mentioned in this chapter. Suggestions for the design of separated laboratories are available from the manufacturers of equipment, such as spectrographic equipment.

6.3 *Safe handling of chemicals and disposal of laboratory wastes*

Many chemicals used routinely in the laboratory for sediment analysis are corrosive, flammable, or toxic. The first step to take when working with chemicals is to find out as much information as possible about their hazardous, handling, and storage requirements, etc. In general, laboratory chemicals should be stored under cool and dry conditions. Explosive and flammable substances must always be stored separately. Certain substances must always be isolated from others (Table 6-1) and should be protected from accidental mixing. Quantities of the chemicals most frequently used and

Table 6-1 Examples of Hazard Ratings and Storage Precautions for Chemical Substances Commonly Used in the Laboratory for Sediment Analysis

Chemical	Hazard rate	Health fire	To be kept separate from
Acetic acid	2	2	Nitric acid, peroxides, and permanganate
Acetone	1	3	Nitric and sulfuric acid
Cyanide	4	4	Acids
Ethyl ether	1	4	Sulfuric acid, sodium
Hydrochloric acid	3	0	
Hydrofluoric acid	4	0	Ammonia, aqueous or anhydrous
Hydrogen peroxide	2	0	Metals, flammable organic compounds
Mercury	4	2	Acetylene, ammonia
Nitric acid	2	0	Acetic acid, acetone, hydrogen sulfide, flammable liquids
Perchloric acid	2	4	Alcohol, paper, oils, wood, acetic anhydride, bismuth
Sulfuric acid	3	0	Potassium, chlorates, permanganate

Note: Hazard ratings from 0 (= no unusual hazard) to 4 (= extreme hazard).
Modified from Steere, 1971; Pipitone, 1984.

generally stored at the laboratory bench, in cupboards, or in cabinets should be limited to the minimum necessary.

Among the strong acids most commonly used for sediment analysis are sulfuric, nitric, hydrochloric, glacial acetic, and hydrofluoric. These concentrated acids are corrosive to the skin; destroy paper, wood, and cloth; and attack most metals. Hot acid solutions in beakers should be handled with beaker tongs. The vapors are poisonous, and a 2% solution of sodium hypochlorite is recommended for swabbing fuming nitric acid splashes (Freeman and Whitehead, 1982). Hydrofluoric acid can cause serious burns which may be painless at first. A common procedure for all sediment laboratories is the acid cleaning of glassware. Nitric and sulfuric acids are among the acids generally used. This routine procedure represents one of the major sources of job hazards in sediment laboratories. Several simple precautions, such as making sure that the person washing the glassware always wears safety goggles, rubber gloves, and a rubber apron, would make this activity much safer. It is desirable to keep a saturated Na_2CO_3 solution in the laboratory to wash the skin should it come in contact with acid.

The most frequently used strong alkalis in sediment analysis are caustic soda, lime, and sodium peroxide. Hot water must not be added to a caustic alkali to dissolve it, because the heat developed might force the liquid out of the reaction vessel (Freeman and Whitehead, 1982). Goggles and gloves must always be worn when handling caustic alkalis.

6.3.1 *Perchloric acid*

Perchloric acid ($HClO_4$), an extremely strong and active oxidizing agent, is often used with a mixture of other acids in the extraction of sediments for the determination of major and trace elements. Perchloric acid, when mixed with easily oxidizable organic or inorganic matter, can produce explosions. The approximate organic content of any sediment should be determined before treating it with perchloric acid. For safety reasons, sediment samples in wet digestions should first be treated with nitric acid to destroy easily oxidizable organic matter before adding perchloric acid. Steere (1971) listed the following hazards of perchloric acid:

- When cold, its properties are those of a strong acid, but when hot, it acts as a strong oxidizing agent.
- It is colorless and will produce severe burns when in contact with the skin, eyes, or respiratory tract.
- It can cause violent explosions if misused or in concentrations greater than normal commercial strength (72%).
- Anhydrous perchloric acid is unstable even at room temperature and ultimately decomposes spontaneously, with a violent explosion.

Fume hoods and ducts used for evaporation of perchloric acid from sediment samples should be made of chemically inert materials such as stainless steel and kept clean and free from combustible materials. Organic chemicals should not be used in hoods designated for perchloric acid digestions. A water spray is desirable for washing down the fume hood after using perchloric acid, to eliminate the buildup of perchlorates. To drain all of the water to the sewer in a satisfactory way during normal wash-down operations, it is important that the exhaust duct go straight up through the building with no horizontal runs (Diberardinis et al., 1987). Any spilled perchloric acid should be removed immediately and the area thoroughly washed with large quantities of water. Any accidental spill of perchloric acid must be neutralized immediately with sodium carbonate and the area washed with large quantities of water (Freeman and Whitehead, 1982). Contact of perchloric acid solution with strong dehydrating agents such as concentrated sulfuric acid may result in the formation of explosive anhydrous perchloric acid. Rubber gloves and goggles should always be worn when treating sediment with perchloric acid. The quantity of perchloric acid in the laboratory should be maintained at a minimum, approximately one reagent bottle (450 g) per fume hood. In summary, perchloric acid may be safely used, provided its hazardous properties are well known and measures are taken to avoid risks. However, unless it is essential, the use of perchloric acid should be avoided in sediment analysis.

6.3.2 Disposal of laboratory wastes

Laboratory users should seek expert advice before disposing of sediments or chemicals whose exact nature is unknown. The disposal of laboratory wastes is regulated, with greater emphasis on wastes from hazardous sediments. Water-soluble wastes that will be diluted down the drain should be non-toxic and not create flammable vapor conditions or stream pollution problems. All disposal through sewers must be in accordance with local guidelines for stream pollution and the type of material acceptable in the public sewer system. Waste solvents should be stored in properly labeled metal safety cans. Wastes that contain sodium or similar materials should be handled and stored separately from wastes that contain water. Incorrect disposal of flammable liquids represents a fire and explosion hazard. Some chemicals react violently with water and others are corrosive and adversely affect the drainage piping and sewage system. Acid and alkaline materials can be carefully neutralized before final disposal. Small quantities of flammable wastes can be burned in shallow metal containers if it has been determined that the vapors will not create problems (Steere, 1971). Because each material is an individual case, consulting the numerous publications that deal with the disposal of dangerous materials (e.g., Gaston, 1970; Steere, 1971; Armour et al., 1987; Lunn and Sansone, 1990; Simmons, 1991) is recommended. It is important to stress that very serious personal hazards and environmental problems may be created by the wastes from a small laboratory.

6.4 Safety in the laboratory for handling highly contaminated sediments

The nature of the sediment will determine whether special clothing or protective equipment, such as respirators, is required. In general, procedures and precautions recommended for working with solid hazardous wastes are valid for highly contaminated sediments (Harris et al., 1984; Diberardinis et al., 1987; Rump and Krist, 1988; Fuentes and Simmons, 1991). The most essential part of the safety program for dealing with hazardous or heavily contaminated sediments is a thorough and effective personnel training program. In addition, safety-oriented discussions or meetings should be held to deal with specific hazards for each particular contaminated sediment being handled in the laboratory.

Toxic substances in sediments can be inhaled, absorbed through the skin, or ingested. Special care should be taken to avoid direct skin contact with contaminated sediments. Precautions should be taken to prevent inhalation of vapors, mainly when working with dry sediments. The potential exposure to toxic substances can be divided into acute and chronic exposures. Acute exposure occurs when laboratory personnel are exposed for a short time to relatively high concentrations of a substance. Chronic exposure occurs from relatively low concentrations of contaminants over a long period of time.

Small quantities of highly toxic sediments should be handled in a completely tight, ventilated glove box. The glove box should be thoroughly decontaminated before disconnecting the exhaust system to avoid loss of contaminants to the laboratory. Large volumes of highly contaminated sediments should be handled in a positive pressure room devoted solely to the analysis of hazardous materials. The air exhausted from fume hoods and glove boxes should be decontaminated before release into the environment. High-efficiency particulate air (HEPA) filters and activated charcoal are commonly used to remove toxic vapors when toxic aerosols are present. A laboratory that handles hazardous sediments should have a water supply, drains with sediment traps, and, where possible, epoxy-coated benches to facilitate easy cleanup. Use of a disposable bench covering during work is considered an added safety practice (Diberardinis et al., 1987). All cracks in floors, walls, and ceilings should be sealed by epoxy or another resistant sealant. Lighting fixtures must be sealed with similar materials so as to be vapor-proof and waterproof. After handling contaminated sediments, the floor and benches should be washed because dry sediment poses a greater health risk than wet sediment. The storage and disposal areas for contaminated sediments should be as close as possible to the laboratory to minimize transport through other areas in the building. In addition to regular safety equipment, such as an emergency eyewash and shower, a laboratory devoted to hazardous sediments should be equipped with an emergency air supply and an alarm. Hydrogen sulfide gases and vapors of different organic contaminants should be monitored in laboratories that work with highly contaminated sediments. Access to a laboratory that handles highly toxic materials should be restricted, and permanent signs indicating restricted access should be posted on the door.

In addition to the safety equipment and procedures recommended for laboratories conducting sediment analysis, every laboratory devoted to handling highly toxic sediments should be provided with an emergency alarm and a changing room with shower facilities. The laboratory must have two separate entrances/exits — one clean and one dirty — to and from the facility with the shower. The following personnel protective equipment should be worn by anyone working with highly contaminated sediments:

- Full overalls with attached hoods and elasticized wrists and ankles
- Neoprene gloves and rubber boots
- Safety glasses and barrier cream on any exposed skin (i.e., wrists or neck)
- Breathing mask (There are several commercially available respirators with multipurpose filters which retain different hazardous materials, such as dust, vapors of organic compounds, and acids

(Stainbrook and Runkle, 1986). Respirators should be stored outside the laboratory to prevent contamination, and filters need to be changed regularly.)

- Contaminated clothing should be rinsed and disposed of

Sieving and grinding of dry sediments generate dust. Both procedures need to be carried out in well-ventilated areas or fume hoods. If grinding is carried out in a mechanical grinder, the grinder should be connected to a ventilation unit or a fume hood. In addition to sufficient ventilation of the area designated for the sieving and grinding of sediments, a breathing mask, safety glasses, and gloves should be used by the laboratory staff carrying out the sieving and grinding.

Sediments collected in areas adjacent to sewer and municipal and industrial effluent outflows may contain different microorganisms. In addition, some laboratories carry out microbiological assessment of sediments. In such cases, safety requirements for working in microbiological laboratories should be followed. Established microbiological laboratories have their own safety requirements. Basic safety guidelines for laboratories handling microbiological materials can be found in different publications (e.g., Reitman and Wedum, 1956).

Laboratory users should not be alone in the laboratory when working with highly toxic sediments. Working alone can be critical in case of an accident or overexposure. If it is not feasible to have a second person in the laboratory, extra precautions should be taken, such as having an audible monitoring system that will allow another person to hear any loud noise or a call for help. Several check procedures, such as telephone checks or personal checks by co-workers, are commonly used for people working alone with hazardous materials. However, practicing the "buddy" system (always two people in the laboratory) is strongly recommended.

References

Armour, M.A., Browne, L.M., and Weir, G.L., *Hazardous Chemicals. Information and Disposal Guide*, 3rd ed., University of Alberta, Edmonton, 1987.

Ashbrook, P.C. and Renfrew, M.M., *Safe Laboratories. Principles and Practices for Design and Remodelling*, Lewis Publishers, Boca Raton, FL, 1991, 166.

Brugger, J., Selection, effectiveness, handling and regeneration of sorbents in the clean up of hazardous material spills, in Proc. 1980 National Conference on Control of Hazardous Chemical Spills, Louisville, KY, May 1980, 92.

Diberardinis, L.J., Baum, J., First, M.W., Gatwood, G.T., Groden, E., and Seth, A.K., *Guidelines for Laboratory Design: Health and Safety Considerations*, John Wiley & Sons, New York, 1987, 285.

Fawcett, H.H., *Hazardous and Toxic Materials — Safe Handling and Disposal*, 2nd ed., John Wiley & Sons, New York, 1988, 514.

Freeman, N.T. and Whitehead, J., *Introduction to Safety in the Chemical Laboratory*, Academic Press, New York, 1982, 244.

Fuentes, A. and Simmons, M., Sampling and analysis of hazardous wastes for toxic pollutants, in *Hazardous Waste Measurements*, Simmons, M.S., Ed., Lewis Publishers, Boca Raton, FL, 1991, 17.

Furr, A.K., *CRC Handbook of Laboratory Safety*, CRC Press, Boca Raton, FL, 1995, 816.

Gaston, P.J., *The Care, Handling and Disposal of Dangerous Chemicals*, The Institute of Science Technology, Northern Publishers (Aberdeen) Ltd., Aberdeen, U.K., 1970.

Harris, J.C., Larsen, D.J., Rechsteiner, C.E., and Thrun, K.E., Sampling and Analysis Methods for Hazardous Waste Combustion, EPA-600/8-84-002, PB84-155845, February 1984.

Hedberg, D., Chemical spills in the storeroom, in Symp. Prudent Practices for the Safe Storage of Chemicals, 187th National Meeting, Division of Chemical Health and Safety, American Chemical Society, Kansas City, MO, September 1982.

Kanare, H.M., *Writing the Laboratory Notebook*, American Chemical Society, Washington, D.C., 1985, 145.

Lunn, G. and Sansone, E.B., *Destruction of Hazardous Chemicals in the Laboratory*, John Wiley & Sons, New York, 1990, 271.

Mudroch, A. and Azcue, J.M., *Manual of Aquatic Sediment Sampling*, Lewis Publishers, Boca Raton, FL, 1995, 219.

Pipitone, D.A., *Safe Storage of Laboratory Chemicals*, John Wiley & Sons, New York, 1984, 280.

Reitman, M. and Wedum, A.G., Microbiological safety, *Public Health J.*, 71, 659, 1956.

Rump, H.H. and Krist, H., *Laboratory Manual for the Examination of Water, Waste Water and Soil*, VCH, Germany, 1988, 190.

Simmons, M.S., *Hazardous Waste Measurements*, Lewis Publishers, Boca Raton, FL, 1991, 315.

Stainbrook, B.W. and Runkle, R.S., Personal protective equipment, in *Laboratory Safety: Principles and Practices*, Miller, B.M., Ed., American Society for Microbiology, Washington, D.C., 1986, 372.

Steere, N.V., *CRC Handbook of Laboratory Safety*, CRC Press, Boca Raton, FL, 1971, 854.

Index

A

B

C

D

H

I

L

M

T

U

V

W

X

Z